Robert Wieser

Domain wall dynamics in quasi one-dimensional nanostructures

Robert Wieser

Domain wall dynamics in quasi one-dimensional nanostructures

Südwestdeutscher Verlag für Hochschulschriften

Impressum / Imprint
Bibliografische Information der Deutschen Nationalbibliothek: Die Deutsche Nationalbibliothek verzeichnet diese Publikation in der Deutschen Nationalbibliografie; detaillierte bibliografische Daten sind im Internet über http://dnb.d-nb.de abrufbar.
Alle in diesem Buch genannten Marken und Produktnamen unterliegen warenzeichen-, marken- oder patentrechtlichem Schutz bzw. sind Warenzeichen oder eingetragene Warenzeichen der jeweiligen Inhaber. Die Wiedergabe von Marken, Produktnamen, Gebrauchsnamen, Handelsnamen, Warenbezeichnungen u.s.w. in diesem Werk berechtigt auch ohne besondere Kennzeichnung nicht zu der Annahme, dass solche Namen im Sinne der Warenzeichen- und Markenschutzgesetzgebung als frei zu betrachten wären und daher von jedermann benutzt werden dürften.

Bibliographic information published by the Deutsche Nationalbibliothek: The Deutsche Nationalbibliothek lists this publication in the Deutsche Nationalbibliografie; detailed bibliographic data are available in the Internet at http://dnb.d-nb.de.
Any brand names and product names mentioned in this book are subject to trademark, brand or patent protection and are trademarks or registered trademarks of their respective holders. The use of brand names, product names, common names, trade names, product descriptions etc. even without a particular marking in this work is in no way to be construed to mean that such names may be regarded as unrestricted in respect of trademark and brand protection legislation and could thus be used by anyone.

Coverbild / Cover image: www.ingimage.com

Verlag / Publisher:
Südwestdeutscher Verlag für Hochschulschriften
ist ein Imprint der / is a trademark of
OmniScriptum GmbH & Co. KG
Heinrich-Böcking-Str. 6-8, 66121 Saarbrücken, Deutschland / Germany
Email: info@svh-verlag.de

Herstellung: siehe letzte Seite /
Printed at: see last page
ISBN: 978-3-8381-5039-0

Zugl. / Approved by: Hamburg, U, Habil., 2014

Copyright © 2015 OmniScriptum GmbH & Co. KG
Alle Rechte vorbehalten. / All rights reserved. Saarbrücken 2015

Contents

1	**Introduction**	**1**
2	**The Heisenberg model**	**3**
3	**The Equations of motion**	**11**
	3.1 Derivation of the Landau-Lifshitz equation	11
	3.2 The Gilbert respectively Landau-Lifshitz-Gilbert equation	25
	3.3 The Landau-Lifshitz-Gilbert-Slonczewski equation	33
	3.4 Numerical method to solve the equation of motion	37
4	**Field and current driven domain wall motion**	**41**
	4.1 Static ferromagnetic domain walls	41
	4.2 Analytical considerations of the dynamics: the q-ϕ model	47
	4.3 Numerical investigation of field driven domain walls	55
	4.4 Combined spin and molecular dynamics study of field driven domain wall motion	65
	4.5 Numerical investigation of current driven domain walls	71
	4.6 Influence of the DM interaction on the domain wall motion	80
	4.7 DM vector perpendicular to the easy axis anisotropy directions	88
	4.7.1 Direct Reversal: DM vector parallel to K_h	89
	4.7.2 Direct Reversal: DM vector perpendicular to K_h	91
	4.7.3 Precessional motion	91
5	**Indirect manipulation of antiferromagnetic domain walls**	**97**
	5.1 Antiferromagnetic domain wall motion in Exchange Bias systems	97
	5.2 Antiferromagnetic domain walls manipulated with a SP-STM	102
6	**Summary**	**115**
A	**Appendix**	**117**
	A.1 Classical elliptical ferromagnetic spin waves (solving the Landau-Lifshitz equation)	117
	A.2 Derivation of the Sine-Gordon equation	118

Contents

Literaturverzeichnis   **121**

1 Introduction

The observation of magnetic nanostructures is a highly topical field of research in recent years. Due to new developments regarding their controlled fabrication and characterization these structures play an important role for basic research as well as for applications in the area of information technology.

Domain walls are good candidates for several applications. Therefore, domain wall dynamics has extensively studied during the last years. The idea behind is to construct logic and storage devices based on domain wall motion. To realize this idea it is necessary to get a complete understanding about domain walls and their dynamics. The main questions are: How can we realize domain wall motion? How fast are the domain walls? Can we control the velocity and how? And at the end, is it possible to make the domain wall faster without loosing the stability? To answer all these questions it is important to perform analytical calculations as well as numerical simulations. The simulations are needed because the analytical calculations mostly base on approximations.

This thesis provides an overview about the dynamics of transverse domain walls. However, in most cases the results can be extended to more complex domain walls.

The manuscript is organized as follows: the second chapter describes the underlying Heisenberg model. It will be shown that this model can be used over a wide range in magnetism starting from a single atom and goes up to macroscopic systems which can be described within the framework of micromagnetism. Depending on the length scale of the system the Heisenberg model has to be the quantum mechanical or the classical Heisenberg model. According to these models and to describe the dynamics we have to deal with different equations of motion. The final goal is the description of the domain wall dynamics will be in the framework of the classical Heisenberg model respectively in the framework of micromagnetism. The underlying equation of motion is the Landau-Lifshitz-Gilbert equation. Originally, the Landau-Lifshitz-Gilbert equation has been introduced phenomenologically. In the second chapter it will be shown that this equation can be derived starting with the dynamical description within the quantum mechanical Schrödinger picture.

The third chapter describes the field and current induced domain wall dynamics starting with the analytical θ-q-model. In the second step these analytical results will be compared with numerical simulations to give a complete picture.

1 Introduction

The fourth chapter describes two alternative ways to manipulate antiferromagnetic domain walls. The first example can be found in Exchange Bias systems. Due to the interlayer coupling between ferromagnetic layer and antiferromagnetic layer also the ferromagnetic and antiferromagnetic domain walls couple. This fact can be used to manipulate the antiferromagnetic domain wall with aid of the ferromagnetic domain wall. The proposal here is to use a conducting ferromagnetic layer and an insulating antiferromagnet. The ferromagnetic domain will be driven by current and the ferromagnetic domain wall pushes the antiferromagnetic domain wall.

The second proposal is to manipulate the antiferromagnetic domain walls with aid of a spin-polarized scanning tunneling microscope. It will be shown that this is possible. For the antiferromagnetic domain wall the description is quite complicated. Therefore, the description will be given for ferromagnetic domain walls. In this case the situation becomes easier and can be explained with some simple rules which can be used also for the antiferromagnetic domain walls.

The thesis ends with a summary and an appendix.

2 The Heisenberg model

Magnetism is the physical phenomenon which is expressed by forces excerted by magnets on other magnets. In nature, we find different types of magnetism like paramagnetism, ferromagnetism, or antiferromagnetism [1, 2, 3]. Independent of the type of magnetism the origin can be seen in uncompensated spins at the core of the atoms or in the electronic bandstructure in metals. Local uncompensated spins lead to local magnetic moments (see right path of Fig. 2.1) which are well described by the quantum mechanical Heisenberg model, which becomes classical in the limit of a huge spin $S \to \infty$.

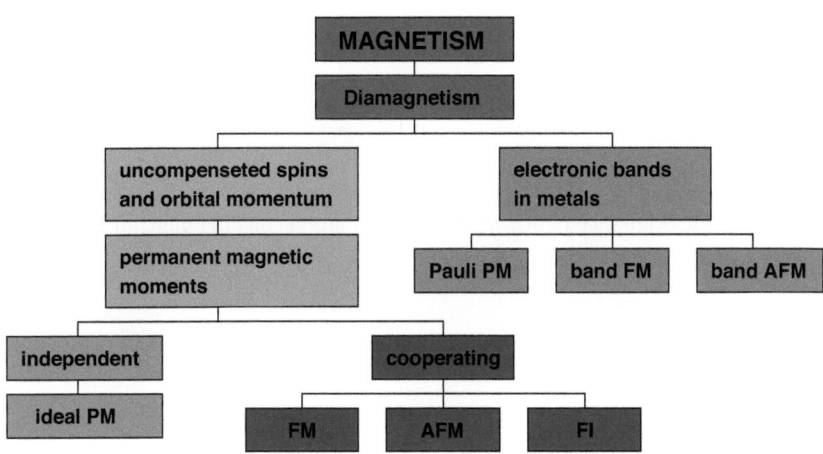

Figure 2.1: Hierarchy of magnetism: PM = paramagnetism, FM = ferromagnetism, AFM = antiferromagnetism, and FI = ferrimagnetism

Originally, the Heisenberg model only describes the exchange interaction between two localized magnetic moments (spins) $\hat{\mathbf{S}}_n$ and $\hat{\mathbf{S}}_m$:

$$\hat{\mathcal{H}} = -\sum_{nm} \frac{J_{nm}}{\hbar^2} \hat{\mathbf{S}}_n \cdot \hat{\mathbf{S}}_m \qquad (2.1)$$

The $\hat{\mathbf{S}}_n = (\hat{S}_n^x, \hat{S}_n^y, \hat{S}_n^z)$ are quantum mechanical spin operators which become in the classical limit ($S \to \infty$, $\hbar \to 0$) three dimensional spin vectors $\mathbf{S}_n = (S_n^x, S_n^y, S_n^z)$. These vectors are mostly assumed to be normalized $\mathbf{S}_{n,m} = \boldsymbol{\mu}/\mu_S$. In this limit the Heisenberg model is given by the following Hamilton function:

$$\mathcal{H} = -\sum_{nm} J_{nm} \mathbf{S}_n \cdot \mathbf{S}_m \qquad (2.2)$$

In the quantum mechanical as well as in the classical limit J_{nm} is the exchange constant which leads to a parallel alignment of the spins (ferromagnetism) if $J_{nm} > 0$ or to an antiparallel alignment (antiferromagnetism) if $J_{nm} < 0$. The exchange interaction can be long ranged (e.g., up to the 10th neighbor spin), however in most cases the exchange interaction can be restricted to the first nearest neighbor. In this case we can skip the index $J_{nm} \to J$ and write:

$$\hat{\mathcal{H}} = -\frac{J}{\hbar^2} \sum_{\langle nm \rangle} \hat{\mathbf{S}}_n \cdot \hat{\mathbf{S}}_m \qquad (2.3)$$

The $\langle \ldots \rangle$ in the index indicates that in this case the summation is over nearest neighbors only.

In the following we want to soften the original definition of the Heisenberg model and allow the occurance of other energy contributions like the uniaxial anisotropy (easy axis in z-direction):

$$\hat{\mathcal{H}}_{D_z} = -\frac{D_z}{\hbar^2} \sum_n (\hat{S}_n^z)^2 \,, \qquad (2.4)$$

the dipole-dipole interaction:

$$\hat{\mathcal{H}}_\omega = -\frac{\omega}{\hbar^2} \sum_{n<m} \frac{3(\hat{\mathbf{S}}_n \cdot \mathbf{e}_{nm})(\hat{\mathbf{S}}_m \cdot \mathbf{e}_{nm}) - \hat{\mathbf{S}}_n \cdot \hat{\mathbf{S}}_m}{r_{nm}^3} \,, \qquad (2.5)$$

or the influence of an external field $\mathbf{B}$ (Zeeman term):

$$\hat{\mathcal{H}}_B = -\frac{g\mu_B}{\hbar} \sum_n \mathbf{B} \cdot \hat{\mathbf{S}}_n \,. \qquad (2.6)$$

The corresponding classical Hamilton functions $\mathcal{H}$ can be derived by replacing the spin operators $\hat{\mathbf{S}}_n$ by classical spins $\mathbf{S}_n$ and by setting $\hbar = 1$.

So far, the description corresponds to the left side (path) of Fig. 2.1. Now, the description shall focus on the second origin of magnetism (right path): The electronic bandstructure leads to itinerant magnetism (ferromagnetism or antiferromagnetism) as well as to the Pauli paramagnetism. The itinerant magnetism can be treated on different length scales: the electronic level, the atomic level, and the microscopic level (micromagnetism).

On the electronic level we need a quantum mechanical description, like the time dependent density functional theory (t-DFT) [4] or a calculation using the Hartree-Fock method [5]. In all these cases, the magnetism is described by the spin density coming from the electrons (see Fig. 2.2).

The highest magnitude of this spin density is located near the core of the atoms [6, 7]. Therefore, we can go over to approximate this by an atomic level description. Here, we assume the magnetism caused by discrete magnetic moments located at the position of the atom. The energy of these magnetic moments is well described by the classical Heisenberg model if we assume a constant absolute value of the magnetic moment. This means that the direction of the magnetic moment can change but not the value. However, this means that we assume that the magnitude of the spin density does not change.

If we further assume that the magnetization is nearly constant over a range of several atoms we can understand the magnetic moments of a certain number of atoms as a larger magnetic moment (macrospin). Under this assumption we are at the microscopic level. The energy of the macrospins can be described similar to the magnetic moments of the atoms with the classical Heisenberg model. This way of description is called discrete micromagnetism and can be used to describe magnetization dynamics on the micrometer scale.

The description using the discrete Heisenberg model is very effective for computer simulations caused by the fact that our computers deal with discrete numbers and not with continuous functions. However, discrete sums are difficult to handle analytically. Therefore, it is more effective to describe the magnetization by a continuum of a nearly constant magnetization density $\mathbf{M}_n \to \mathbf{M}(\mathbf{r})$. In this case the sums become integrals and the description is called micromagnetism.

The goal now is to show how to transform the classical Heisenberg Hamiltonian $\mathcal{H}$ into the according energy E of the micromagnetic description.

In the case of the anisotropy and the Zeeman term the transformation can be easily done just by replacing the sum by an integral over the space and the discrete magnetization $\mathbf{M}_n = (1/N \sum_{i=1}^{N} \mathbf{S}_i)_n$ by the continuous magnetization $\mathbf{M}(\mathbf{r})$. This results in the following two energies:

$$E_B = \int d^3\mathbf{r}\, \mathbf{B} \cdot \mathbf{M}(\mathbf{r}) \qquad (2.7)$$

2 The Heisenberg model

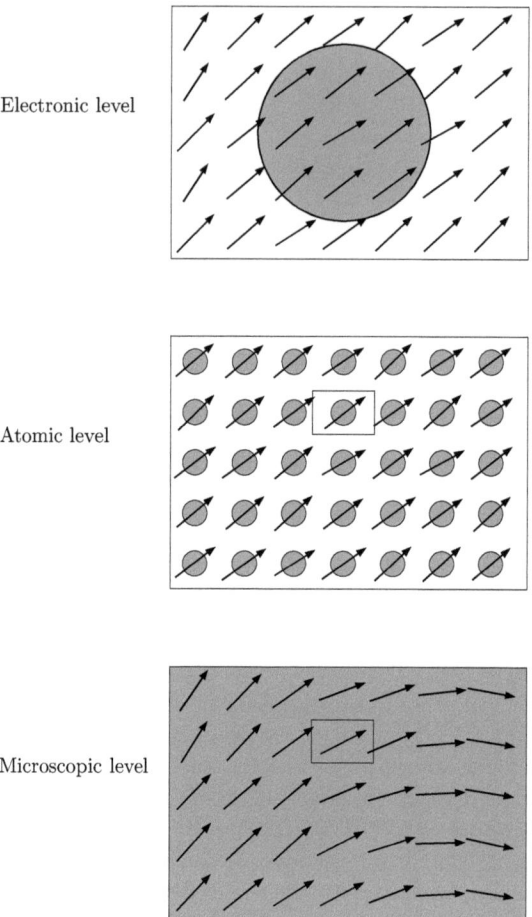

Figure 2.2: Magnetization on different length scales: electronic level, atomic level and the microscopic level. The light blue areas in all figures illustrate the areas of large magnitude of the magnetization. On the electronic level the magnetization is given by the spin density. The highest magnitude appears near the atomic core (light blue area). On the atomic level, the magnetization is described by discrete spins located at the atomic positions. On the microscopic level the magnetization is represented by macrospins which are the sum over the atomic spins in a certain area.

and

$$E_{D_z} = \int d^3\mathbf{r}\, K M_z^2(\mathbf{r})\ . \tag{2.8}$$

The exchange interaction contains two magnetizations $\mathbf{M}_n$ and $\mathbf{M}_m$ which belong to two different lattice sites. Therefore, we cannot transform this Hamiltonian in the same manner. In this case we would end up with $\mathbf{M}(\mathbf{r}_n)$ and $\mathbf{M}(\mathbf{r}_m)$ where the $\mathbf{r}_{n,m}$ are undefined. In this case we need to know how the magnetization changes with $\mathbf{r}$.

Starting point is the Hamilton function for the exchange interaction:

$$\mathcal{H}_J = -\frac{J}{2} \sum_{\langle ij \rangle} \mathbf{M}_i \cdot \mathbf{M}_j \tag{2.9}$$

Here, the factor 1/2 has been introduced because of the double summation.

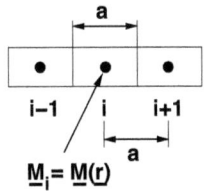

Figure 2.3: Scheme of the magnetic lattice

Now, there are different ways to get the micromagnetic exchange interaction. The most common way starts with the assumption that we have a continuous magnetization:

$$\mathbf{M}_i \approx \mathbf{M}(\mathbf{r}) \tag{2.10}$$

and the assumption that $\mathbf{M}_{i\pm1}$ can be expressed as a Taylor expansion around $\mathbf{M}_i$:

$$\mathbf{M}_{i\pm1} \approx \mathbf{M}(\mathbf{r}) \pm a\frac{d\mathbf{M}}{dr} + \frac{a^2}{2}\frac{d^2\mathbf{M}}{dr^2}\ , \tag{2.11}$$

with a the lattice constant (see figure 2). This means that we assume that the magnetization changes slowly with $\mathbf{r}$ resp. i.

Using these expressions, the summands of $\mathcal{H}_J$ containing i become:

$$-\frac{J}{2}\left[(\mathbf{M}_{i-1}\cdot\mathbf{M}_i)+(\mathbf{M}_i\cdot\mathbf{M}_{i+1})\right]$$

$$\approx -\frac{J}{2}\left[\left(\mathbf{M}(\mathbf{r})-a\frac{d\mathbf{M}}{dr}+\frac{a^2}{2}\frac{d^2\mathbf{M}}{dr^2}\right)\cdot\mathbf{M}(\mathbf{r})+\mathbf{M}(\mathbf{r})\cdot\left(\mathbf{M}(\mathbf{r})+a\frac{d\mathbf{M}}{dr}+\frac{a^2}{2}\frac{d^2\mathbf{M}}{dr^2}\right)\right]$$

$$= -JM^2 - \frac{Ja^2}{2}\mathbf{M}(\mathbf{r})\cdot\frac{d^2\mathbf{M}}{dr^2} \qquad (2.12)$$

Now, we have:

$$2\mathbf{M}\frac{d^2\mathbf{M}}{dr^2} = \frac{d}{dr}\left(2\mathbf{M}\frac{d\mathbf{M}}{dr}\right) - 2\left(\frac{d\mathbf{M}}{dr}\right)^2 \qquad (2.13)$$

However, the second first term on the right hand side is zero, because $\mathbf{M}\perp d\mathbf{M}/dr$. Therefore, Eq. (2.12) becomes the exchange energy density:

$$\mathcal{E}_J = -JM^2 + \frac{Ja^2}{2}\left(\frac{d\mathbf{M}}{dr}\right)^2. \qquad (2.14)$$

The first term is just a change of the zero point of energy and can be skipped. Then, with $A = J/2a$ and $dV = a^3$ we get the exchange energy corresponding to Eq. 2.9:

$$E_J = \int \mathcal{E}_J\, dV = A\int\left(\frac{d\mathbf{M}}{dr}\right)^2 dV. \qquad (2.15)$$

The micromagnetic exchange energy in spherical coordinates can be obtained by:

$$\mathbf{M} = M\mathbf{e}_M = \begin{pmatrix} M\sin\theta\cos\phi \\ M\sin\theta\sin\phi \\ M\cos\theta \end{pmatrix}, \qquad (2.16)$$

and

$$\frac{d\mathbf{M}}{dr} = \frac{dM}{dr}\mathbf{e}_M + M\frac{d\mathbf{e}_M}{dr} = \frac{dM}{dr}\mathbf{e}_M + M\frac{d\theta}{dr}\mathbf{e}_\theta + M\sin\theta\frac{d\phi}{dr}\mathbf{e}_\phi, \qquad (2.17)$$

where

$$\mathbf{e}_M = \begin{pmatrix} \sin\theta\cos\phi \\ \sin\theta\sin\phi \\ \cos\theta \end{pmatrix} \quad \mathbf{e}_\theta = \begin{pmatrix} \cos\theta\cos\phi \\ \cos\theta\sin\phi \\ -\sin\theta \end{pmatrix} \quad \mathbf{e}_\phi = \begin{pmatrix} -\sin\phi \\ \cos\phi \\ 0 \end{pmatrix} \qquad (2.18)$$

are the three orthogonal unit vectors with $\mathbf{e}_\phi\times\mathbf{e}_M = \mathbf{e}_\theta$ respectively, $\mathbf{e}_M\times\mathbf{e}_\phi = -\mathbf{e}_\theta$ and cyclic commutations.

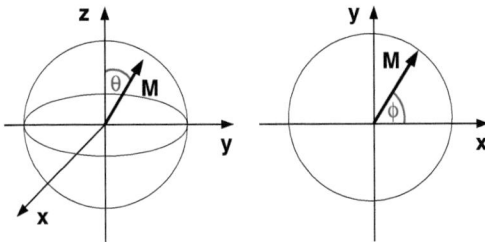

Figure 2.4: Spherical coordinates

With $M = 1 = const.$ we find

$$\frac{d\mathbf{M}}{d\mathbf{r}} = M\frac{d\mathbf{e}_M}{d\mathbf{r}} = \frac{d\theta}{d\mathbf{r}}\mathbf{e}_\theta + \sin\theta\frac{d\phi}{d\mathbf{r}}\mathbf{e}_\phi = \begin{pmatrix} \frac{d\theta}{d\mathbf{r}} \\ \sin\theta\frac{d\phi}{d\mathbf{r}} \end{pmatrix} . \tag{2.19}$$

and finally

$$E_\text{J} = A \int \left[\left(\frac{d\theta}{d\mathbf{r}}\right)^2 + \sin^2\theta \left(\frac{d\phi}{d\mathbf{r}}\right)^2 \right] dV . \tag{2.20}$$

The way to deal with the dipole-dipole interaction in the continuum limit is the same as in classical electrodynamics. We have to solve the Maxwell equations [8] e.g.:

$$\text{div}\mathbf{B} = \mu_0 \text{div}\left(\mathbf{H} + \mathbf{M}\right) = 0 . \tag{2.21}$$

So far, the Heisenberg model has been introduced. However, the Heisenberg model just describes the energy contributions of the system. To describe dynamics we need an equation of motion.

2 The Heisenberg model

3 The Equations of motion

3.1 Derivation of the Landau-Lifshitz equation

This chapter, describes three simple ways to derive the Landau-Lifshitz equation starting from the quantum mechanical time evolution. Such a derivation provides a deeper understanding of the underlying mathematics and the connection between quantum mechanics and classical physics.

The complete derivation starts with the quantum mechanical time evolution of the state $|\psi(t)\rangle$:

$$|\psi(t+\Delta t)\rangle = \hat{U}(t+\Delta t, t)|\psi(t)\rangle. \tag{3.1}$$

Under the assumption of a small time step Δt we can expand the time evolution operator $\hat{U}(t+\Delta t, t) \approx (\hat{1} - i\hat{\mathcal{H}}\Delta t/\hbar) + \mathcal{O}(\Delta t^2)$:

$$|\psi^{(1)}(t+\Delta t)\rangle \approx \left(\hat{1} - \frac{i\hat{\mathcal{H}}\Delta t}{\hbar}\right)|\psi(t)\rangle \tag{3.2}$$

If we further assume that we have a non-Hermitian Hamilton operator: $\hat{\mathcal{H}}^+ \neq \hat{\mathcal{H}}$ we get for the corresponding bra vector:

$$\langle\psi^{(1)}(t+\Delta t)| \approx \langle\psi(t)|\left(1 + \frac{i\hat{\mathcal{H}}^+\Delta t}{\hbar}\right), \tag{3.3}$$

and for the norm:

$$\begin{aligned}
n^2 &= \langle\psi^{(1)}(t+\Delta t)|\psi^{(1)}(t+\Delta t)\rangle \\
&= \langle\psi(t)|\left(1 + \frac{i\hat{\mathcal{H}}^+\Delta t}{\hbar}\right)\left(1 - \frac{i\hat{\mathcal{H}}\Delta t}{\hbar}\right)|\psi(t)\rangle \\
&= 1 - \frac{i}{\hbar}\Delta t\langle\psi(t)|\hat{\mathcal{H}} - \hat{\mathcal{H}}^+|\psi(t)\rangle = 1 - r.
\end{aligned} \tag{3.4}$$

We are now looking for a normalized wave function and make the ansatz:

$$|\psi(t+\Delta t)\rangle = \frac{|\psi^{(1)}(t+\Delta t)\rangle}{\sqrt{1-r}} \tag{3.5}$$

3 The Equations of motion

Then, Eq. (3.2) can be rewritten

$$\frac{|\psi^{(1)}(t+\Delta t)\rangle - |\psi(t)\rangle}{\Delta t} = -\frac{i}{\hbar}\hat{\mathcal{H}}|\psi(t)\rangle \,, \tag{3.6}$$

and with Eq. (3.5)

$$\frac{|\psi(t+\Delta t)\rangle\sqrt{1-r} - |\psi(t)\rangle}{\Delta t} = -\frac{i}{\hbar}\hat{\mathcal{H}}|\psi(t)\rangle \,. \tag{3.7}$$

Furthermore, with the Taylor expansion: $\sqrt{1-r} \approx 1 - \frac{1}{2}r$ we get:

$$\frac{|\psi(t+\Delta t)\rangle - |\psi(t)\rangle}{\Delta t} - \frac{1}{2}\frac{r}{\Delta t}|\psi(t+\Delta t)\rangle = -\frac{i}{\hbar}\hat{\mathcal{H}}|\psi(t)\rangle \tag{3.8}$$

where r is given by Eq. (3.4). In the limit, $\Delta t \to 0$ the differential quotient becomes a differential operator dt and $|\psi(t+\Delta t)\rangle$ becomes $|\psi(t)\rangle$. Finally, we get the following modified time dependent Schrödinger equation:

$$i\hbar\frac{d}{dt}|\psi(t)\rangle = \left(\hat{\mathcal{H}} + \langle\psi(t)|\frac{\hat{\mathcal{H}}^+ - \hat{\mathcal{H}}}{2}|\psi(t)\rangle\right)|\psi(t)\rangle \,. \tag{3.9}$$

This formula is identical with the equation proposed by K. Mølmer et al. [9] for the calculation of Monte Carlo wave functions in quantum optics.

With $\hat{\mathcal{H}} = \hat{H} - i\lambda\hat{\Gamma}$, $\hat{\mathcal{H}}^+ = \hat{H} + i\lambda\hat{\Gamma}$ ($\lambda \in \mathbb{R}_0^+$, $\hat{\Gamma}$ Hermitian), and $\langle\hat{\Gamma}\rangle = \langle\psi(t)|\hat{\Gamma}|\psi(t)\rangle$, Eq. (3.9) becomes:

$$i\hbar\frac{d}{dt}|\psi(t)\rangle = (\hat{H} - i\lambda[\hat{\Gamma} - \langle\hat{\Gamma}\rangle])|\psi(t)\rangle \,. \tag{3.10}$$

N. Gisin [10] has proposed a similar equation, however, with the use of $\hat{\mathcal{H}} = \hat{H} - i\lambda\hat{H}$:

$$i\hbar\frac{d}{dt}|\psi(t)\rangle = (\hat{H} - i\lambda[\hat{H} - \langle\hat{H}\rangle])|\psi(t)\rangle \,. \tag{3.11}$$

This special case of Eq. (3.10) is the quantum mechanical counterpart of the Landau-Lifshitz equation as will be shown below. Please notice $\hat{\mathcal{H}}$ resp. $\hat{H}$ or $\hat{\Gamma}$ itself can be time dependent.

Now, Eq. (3.11) can be rewritten as:

$$\begin{aligned} i\hbar\frac{d}{dt}|\psi\rangle &= \hat{H}|\psi\rangle - i\lambda\left(\hat{H}|\psi\rangle\underbrace{\langle\psi|\psi\rangle}_{=1} - |\psi\rangle\langle\psi|\hat{H}|\psi\rangle\right) \\ &= \hat{H}|\psi\rangle - i\lambda\left(\hat{H}|\psi\rangle\langle\psi| - |\psi\rangle\langle\psi|\hat{H}\right)|\psi\rangle \,, \end{aligned}$$

3.1 Derivation of the Landau-Lifshitz equation

and finally:

$$i\hbar \frac{\mathrm{d}}{\mathrm{d}t}|\psi\rangle = \left(\hat{H} - i\lambda\left[\hat{H}, |\psi\rangle\langle\psi|\right]\right)|\psi\rangle . \tag{3.12}$$

The corresponding conjugate transposed equation is given by:

$$-i\hbar \frac{\mathrm{d}}{\mathrm{d}t}\langle\psi| = \langle\psi|\left(\hat{H} + i\lambda\left[\hat{H}, |\psi\rangle\langle\psi|\right]^{+}\right) . \tag{3.13}$$

Now, the conjugate transposed version of the commutator $\left[\hat{H}, |\psi\rangle\langle\psi|\right]$ will be derived.

We know that $\hat{H}^{+} = \hat{H}$, because $\hat{H}$ is Hermitian. Therefore, we can write:

$$\begin{aligned}
\left[\hat{H}, |\psi\rangle\langle\psi|\right]^{+} &= \left(\underbrace{\langle\psi|\psi\rangle}_{=1}\right)^{+}\hat{H}^{+}|\psi\rangle^{+}\langle\psi|^{+} - |\psi\rangle^{+}\langle\psi|^{+}\hat{H}^{+}\left(\underbrace{\langle\psi|\psi\rangle}_{=1}\right)^{+} \\
&= |\psi\rangle\langle\psi|\hat{H}\underbrace{\langle\psi|\psi\rangle}_{=1} - \underbrace{\langle\psi|\psi\rangle}_{=1}\hat{H}|\psi\rangle\langle\psi| \\
&= \left[|\psi\rangle\langle\psi|, \hat{H}\right] .
\end{aligned} \tag{3.14}$$

This means:

$$\left[\hat{H}, |\psi\rangle\langle\psi|\right]^{+} = -\left[\hat{H}, |\psi\rangle\langle\psi|\right] , \tag{3.15}$$

and therefore the corresponding conjugate transposed equation becomes:

$$-i\hbar \frac{\mathrm{d}}{\mathrm{d}t}\langle\psi| = \langle\psi|\left(\hat{H} - i\lambda\left[\hat{H}, |\psi\rangle\langle\psi|\right]\right) . \tag{3.16}$$

With this equation, we are able to write down the von Neumann equation, corresponding to Eq. (3.11):

$$\begin{aligned}
\frac{\mathrm{d}}{\mathrm{d}t}\left(|\psi\rangle\langle\psi|\right) &= \frac{\mathrm{d}|\psi\rangle}{\mathrm{d}t}\langle\psi| + |\psi\rangle\frac{\mathrm{d}\langle\psi|}{\mathrm{d}t} \\
&= -\frac{i}{\hbar}\left(\hat{H} - i\lambda\left[\hat{H}, |\psi\rangle\langle\psi|\right]\right)|\psi\rangle\langle\psi| + \frac{i}{\hbar}|\psi\rangle\langle\psi|\left(\hat{H} - i\lambda\left[\hat{H}, |\psi\rangle\langle\psi|\right]\right) \\
&= \frac{i}{\hbar}\left(|\psi\rangle\langle\psi|\hat{H} - \hat{H}|\psi\rangle\langle\psi|\right) + \frac{\lambda}{\hbar}\left(|\psi\rangle\langle\psi|\left[\hat{H}, |\psi\rangle\langle\psi|\right] - \left[\hat{H}, |\psi\rangle\langle\psi|\right]|\psi\rangle\langle\psi|\right) \\
&= \frac{i}{\hbar}\left(|\psi\rangle\langle\psi|\hat{H} - \hat{H}|\psi\rangle\langle\psi|\right) - \frac{\lambda}{\hbar}\left(|\psi\rangle\langle\psi|\left[|\psi\rangle\langle\psi|, \hat{H}\right] - \left[|\psi\rangle\langle\psi|, \hat{H}\right]|\psi\rangle\langle\psi|\right) \\
&= \frac{i}{\hbar}\left[|\psi\rangle\langle\psi|, \hat{H}\right] - \frac{\lambda}{\hbar}\left[|\psi\rangle\langle\psi|, \left[|\psi\rangle\langle\psi|, \hat{H}\right]\right] .
\end{aligned}$$

3 The Equations of motion

With $\hat{\rho} = |\psi\rangle\langle\psi|$ we finally obtain the von Neumann equation:

$$\frac{d\hat{\rho}}{dt} = \frac{i}{\hbar}\left[\hat{\rho}, \hat{H}\right] - \frac{\lambda}{\hbar}\left[\hat{\rho}, \left[\hat{\rho}, \hat{H}\right]\right]. \quad (3.17)$$

The common way to derive the undamped Landau-Lifshitz equation is to write down the Heisenberg equation:

$$\frac{d\hat{\mathbf{S}}}{dt} = -\frac{i}{\hbar}\left[\hat{\mathbf{S}}, \hat{H}\right] \quad (3.18)$$

and to show that the commutator $i[\hat{\mathbf{S}}, \hat{H}]$ corresponds to the vector product $\hat{\mathbf{S}} \times \partial\hat{H}/\partial\hat{\mathbf{S}}$. The Landau-Lifshitz equation

$$\frac{d\mathbf{S}}{dt} = -\frac{\gamma}{\mu_S}\mathbf{S} \times \frac{\partial\mathcal{H}}{\partial\mathbf{S}} \quad (3.19)$$

appears in the classical limit $S \to \infty$. Thereby, γ is the gyromagnetic ratio and $\mu_S = |\mathbf{S}|$.

Both, the Heisenberg equation (3.18) and the von Neumann equation (3.17) in the case of no damping ($\lambda = 0$) are similar, except for the minus sign of the right hand side of the von Neumann equation. Furthermore, if we know the von Neumann equation we can immediately write down the Heisenberg equation just by replacing the density operator $\hat{\rho}$ by the spin operator $\hat{\mathbf{S}}$ and by changing the sign in front of the commutator. For the double commutator $[\hat{\rho}, [\hat{\rho}, \hat{H}]]$ the sign changes twice which means that there is no change of the sign.

With this recipe we can write down the Heisenberg equation with an additional damping term as:

$$\frac{d\hat{\mathbf{S}}}{dt} = -\frac{i}{\hbar}\left[\hat{\mathbf{S}}, \hat{H}\right] - \frac{\lambda}{\hbar}\left[\hat{\mathbf{S}}, \left[\hat{\mathbf{S}}, \hat{H}\right]\right]. \quad (3.20)$$

In the limit $S \to \infty$, this equation becomes the Landau-Lifshitz equation:

$$\frac{d\mathbf{S}}{dt} = -\frac{\gamma}{\mu_S}\mathbf{S} \times \mathbf{H}_{\text{eff}} - \frac{\lambda}{\mu_S}\mathbf{S} \times (\mathbf{S} \times \mathbf{H}_{\text{eff}}), \quad (3.21)$$

with

$$\mathbf{H}_{\text{eff}} = -\frac{\partial\mathcal{H}}{\partial\mathbf{S}}. \quad (3.22)$$

Nevertheless, there are two problems: First, the derivation bases on a comparison. And second, it is easy to show that the commutator $i[\hat{\mathbf{S}}, \hat{H}]$ and the classical Poisson bracket $\{\mathbf{S}, \mathcal{H}\}$ corresponds to the vector product $\hat{\mathbf{S}} \times \partial\hat{H}/\partial\hat{\mathbf{S}}$

3.1 Derivation of the Landau-Lifshitz equation

and $\mathbf{S} \times \partial \mathcal{H}/\partial \mathbf{S}$ [11], respectively. This means that the double Poisson bracket $\{\mathbf{S}, \{\mathbf{S}, \mathcal{H}\}\}$ results in $\mathbf{S} \times \left(\frac{\partial}{\partial \mathbf{S}}\left(\mathbf{S} \times \frac{\partial \mathcal{H}}{\partial \mathbf{S}}\right)\right)$. Moreover, while $i[\hat{\mathbf{S}}, \hat{\mathbf{H}}]$ immediately results in the vectorproduct $\hat{\mathbf{S}} \times \partial \hat{\mathbf{H}}/\partial \hat{\mathbf{S}}$ after working out the commutator the result which appears for $[\hat{\mathbf{S}}, [\hat{\mathbf{S}}, \hat{\mathbf{H}}]]$ does not correspond to the double vector product $\hat{\mathbf{S}} \times (\hat{\mathbf{S}} \times \partial \hat{\mathbf{H}}/\partial \hat{\mathbf{S}})$. On the other hand, there is an expression of the Landau-Lifshitz equation using the double Poisson bracket $\{\mathbf{S}, \{\mathbf{S}, \mathcal{H}\}\}$ [12]. However, in this case $\mathbf{S}$ and $\mathcal{H}$ have to be 3×3 skew-symmetric matrices. In conclusion, we can say that the Heisenberg Eq. (3.20) is not correct.

The correct Heisenberg Eq. is given by:

$$\frac{d}{dt}\hat{\mathbf{S}} = -\frac{i}{\hbar}\left[\hat{\mathbf{S}}, \hat{\mathbf{H}}\right] - \frac{\lambda}{\hbar}\left(\left[\hat{\mathbf{S}}, \hat{\mathbf{H}}\right]_+ - 2\langle\hat{\mathbf{H}}\rangle\hat{\mathbf{S}}\right), \tag{3.23}$$

where $[\hat{\mathbf{S}}, \hat{\mathbf{H}}]_+ = \hat{\mathbf{S}}\hat{\mathbf{H}} + \hat{\mathbf{H}}\hat{\mathbf{S}}$ ist the anticommutator. This equation can be derived directly from the definition of the Heisenberg equation

$$i\hbar\frac{d}{dt}\hat{\mathbf{S}} = \left[\hat{\mathbf{S}}, \hat{\mathcal{H}}\right] = \hat{\mathbf{S}}\hat{\mathcal{H}} - \hat{\mathcal{H}}^+\hat{\mathbf{S}}, \tag{3.24}$$

if we assume that the Hamiltonian $\hat{\mathcal{H}}$ is non-Hermitian: $\hat{\mathcal{H}} \neq \hat{\mathcal{H}}^+$, e.g. $\hat{\mathcal{H}} = \hat{\mathbf{H}} - i\lambda\hat{\Gamma}$ (in our case $\hat{\Gamma} = \hat{\mathbf{H}}$) and the norm is not conserved but decaying according to [13]:

$$\hbar\frac{dn^2}{dt} = -2\lambda\langle\psi|\hat{\Gamma}|\psi\rangle. \tag{3.25}$$

Alternatively, this equation can be derived using the von Neumann Eq. (3.17). In this case we concentrate on the time dependence of the expectation value $\langle\hat{\mathbf{S}}\rangle$:

$$\frac{d}{dt}\langle\hat{\mathbf{S}}\rangle = \text{Tr}\left(\frac{d\hat{\rho}}{dt}\hat{\mathbf{S}}\right), \tag{3.26}$$

where $d\hat{\rho}/dt$ is given by the von Neumann equation, which can be written as:

$$\begin{aligned}\frac{d\hat{\rho}}{dt} &= \frac{i}{\hbar}\left[\hat{\rho}, \hat{\mathbf{H}}\right] - \frac{\lambda}{\hbar}\left[\hat{\rho}, \left[\hat{\rho}, \hat{\mathbf{H}}\right]\right] \\ &= \frac{i}{\hbar}\left[\hat{\rho}, \hat{\mathbf{H}}\right] - \frac{\lambda}{\hbar}\left(\hat{\rho}^2\hat{\mathbf{H}} + \hat{\mathbf{H}}\hat{\rho}^2 - 2\hat{\rho}\hat{\mathbf{H}}\hat{\rho}\right) \\ &= \frac{i}{\hbar}\left[\hat{\rho}, \hat{\mathbf{H}}\right] - \frac{\lambda}{\hbar}\left(\left[\hat{\rho}, \hat{\mathbf{H}}\right]_+ + 2\hat{\rho}\hat{\mathbf{H}}\hat{\rho}\right).\end{aligned} \tag{3.27}$$

Here, we use the fact that $\hat{\rho}^2 = \hat{\rho}$.

Consequenty, Eq. (3.26) becomes:

$$\frac{d}{dt}\langle\hat{\mathbf{S}}\rangle = -\frac{i}{\hbar}\langle[\hat{\mathbf{S}}, \hat{\mathbf{H}}]\rangle - \frac{\lambda}{\hbar}\left(\langle[\hat{\mathbf{S}}, \hat{\mathbf{H}}]_+\rangle - 2\langle\hat{\mathbf{H}}\rangle\langle\hat{\mathbf{S}}\rangle\right). \tag{3.28}$$

3 The Equations of motion

Here, we have used the fact that the trace is invariant under cyclic permutations, and that

$$\begin{align}
\text{Tr}\left(\hat{\rho}\hat{H}\hat{\rho}\hat{\mathbf{S}}\right) &= \sum_n \langle n|\psi\rangle\langle\psi|\hat{H}|\psi\rangle\langle\psi|\hat{\mathbf{S}}|n\rangle \\
&= \sum_n \langle\psi|\hat{H}|\psi\rangle\langle\psi|\hat{\mathbf{S}}|n\rangle\langle n|\psi\rangle \\
&= \langle\psi|\hat{H}|\psi\rangle\langle\psi|\hat{\mathbf{S}}|\psi\rangle \\
&= \langle\hat{H}\rangle\langle\hat{\mathbf{S}}\rangle .
\end{align} \tag{3.29}$$

The Heisenberg Eq. (3.23) appears if we replace the expectation values $\langle\hat{\mathbf{S}}\rangle$ by $\hat{\mathbf{S}}$ in Eq. (3.28). However, there is still a problem. The Landau-Lifshitz Eq. (3.21) is supposed to be the classical limit of the Heisenberg Eq. (3.23), but it is hard to see that $-\lambda/\hbar(\langle[\hat{\mathbf{S}},\hat{H}]_+\rangle - 2\langle\hat{H}\rangle\langle\hat{\mathbf{S}}\rangle)$ corresponds to the classical expression $\lambda/\mu_S(\mathbf{S}\times(\mathbf{S}\times\partial\mathcal{H}/\partial\mathbf{S}))$.

But all these problems can be avoided, because it is not necessary to write down the Heisenberg equation to derive the Landau-Lifshitz equation (3.21). The von Neumann equation can be interpreted as the quantum mechanical counterpart of the Landau-Lifshitz equation.

For $S=\frac{1}{2}$ the density operator is given by:

$$\hat{\rho} = \frac{1}{2}\left(\hat{\mathbf{1}} + \langle\hat{\sigma}\rangle\hat{\sigma}\right) . \tag{3.30}$$

The factor $1/2$ is just for the normation because $\text{Tr}\hat{\rho} = 1$. The unity matrix $\hat{\mathbf{1}}$ does commutate with $\hat{H}$, therefore this term can be skipped and $\hat{\rho}$ is equal to the polarization $\langle\hat{\sigma}\rangle\hat{\sigma}$ with the Pauli matrix vector $\hat{\sigma} = (\hat{\sigma}_x, \hat{\sigma}_y, \hat{\sigma}_z)$. Here, σ_η, $\eta \in \{x,y,z\}$ are the Pauli matrices. For general, S the polarization $\langle\hat{\sigma}\rangle\hat{\sigma}$ has to be replaced by $\langle\hat{\mathbf{S}}\rangle\hat{\mathbf{S}}/S$ with the corresponding spin (matrix) vector $\hat{\mathbf{S}} = (\hat{S}_x, \hat{S}_y, \hat{S}_z)$ and $\hat{S}_\eta = \bigotimes_{n=1}^{2S} \sigma_n^\eta, \eta \in \{x,y,z\}$:

$$\hat{\rho} \approx \frac{\langle\hat{\mathbf{S}}\rangle\hat{\mathbf{S}}}{\hbar S} . \tag{3.31}$$

Then, Eq. (3.31) can be written as:

$$\hat{\rho} \approx \mathbf{P}\cdot\hat{\mathbf{S}} = P_x\hat{S}_x + P_y\hat{S}_y + P_z\hat{S}_z , \tag{3.32}$$

with the normalized polarization $\mathbf{P} = \langle\hat{\mathbf{S}}\rangle/\hbar S$.

At this point let us make the assumption that we can write the Hamilton operator as:

$$\hat{H} = -\mathbf{B}\cdot\hat{\mathbf{S}}/\hbar . \tag{3.33}$$

3.1 Derivation of the Landau-Lifshitz equation

B is like an effective field and should be independent of $\hat{\mathbf{S}}$. This is always possible in the case of Hamilton operators linear in $\hat{\mathbf{S}}_n$. Here, **B** is independent of $\hat{\mathbf{S}}_n$, while we are interested in the time dependence of $\langle\hat{\mathbf{S}}_n\rangle$. If **B** is not independent of $\hat{\mathbf{S}}$ (or $\hat{\mathbf{S}}_n$) the following description will not fit and the quantum mechanical description leads to different results compared to the classical Landau-Lifshitz equation. This situation shall be discussed later.

Now, we are able to write down the terms of the von Neumann equation (3.17). Let's start with the first term (without all prefactors):

$$i[\hat{\rho},\hat{H}] = -\frac{i}{\hbar}[P_x\hat{S}_x + P_y\hat{S}_y + P_z\hat{S}_z, B_x\hat{S}_x + B_y\hat{S}_y + B_z\hat{S}_z] \,. \tag{3.34}$$

This can be written as:

$$\begin{aligned}i[\hat{\rho},\hat{H}] = & -\frac{i}{\hbar}(P_xB_x[\hat{S}_x,\hat{S}_x] + P_xB_y[\hat{S}_x,\hat{S}_y] + P_xB_z[\hat{S}_x,\hat{S}_z]\\ & + P_yB_x[\hat{S}_y,\hat{S}_x] + P_yB_y[\hat{S}_y,\hat{S}_y] + P_yB_z[\hat{S}_y,\hat{S}_z]\\ & + P_zB_x[\hat{S}_z,\hat{S}_x] + P_zB_y[\hat{S}_z,\hat{S}_y] + P_zB_z[\hat{S}_z,\hat{S}_z])\,.\end{aligned} \tag{3.35}$$

Working out the commutators ($[\hat{S}_\alpha,\hat{S}_\beta] = i\hbar\epsilon_{\alpha\beta\gamma}\hat{S}_\gamma$, where $\{\alpha,\beta,\gamma\}$ stands for $\{x,y,z\}$ and $\epsilon_{\alpha\beta\gamma}$ is the Levi-Cevita tensor), leads to:

$$i[\hat{\rho},\hat{H}] = P_xB_y\hat{S}_z - P_xB_z\hat{S}_y - P_yB_x\hat{S}_z + P_yB_z\hat{S}_x + P_zB_x\hat{S}_y - P_zB_y\hat{S}_x\,. \tag{3.36}$$

However, this is nothing else than:

$$\begin{pmatrix}P_yB_z - P_zB_y\\ P_zB_x - P_xB_z\\ P_xB_y - P_yB_x\end{pmatrix} = \begin{pmatrix}P_x\\ P_y\\ P_z\end{pmatrix} \times \begin{pmatrix}B_x\\ B_y\\ B_z\end{pmatrix} = \mathbf{P}\times\mathbf{B}\,. \tag{3.37}$$

Here, we have used the fact that the $\hat{S}_\alpha$, $\alpha \in \{x,y,z\}$ give a cartesian coordinate system. Eq. (3.37) describes a precessional motion of the components of the polarization **P** in the effective field **B**. The second term of Eq. (3.17) can be written as:

$$\begin{aligned}\frac{1}{\hbar}[\hat{\rho},[\hat{\rho},\hat{H}]] = & -\frac{i}{\hbar}[P_x\hat{S}_x, P_xB_y\hat{S}_z - P_xB_z\hat{S}_y - P_yB_x\hat{S}_z + P_yB_z\hat{S}_x + P_zB_x\hat{S}_y - P_zB_y\hat{S}_x]\\ & -\frac{i}{\hbar}[P_y\hat{S}_y, P_xB_y\hat{S}_z - P_xB_z\hat{S}_y - P_yB_x\hat{S}_z + P_yB_z\hat{S}_x + P_zB_x\hat{S}_y - P_zB_y\hat{S}_x]\\ & -\frac{i}{\hbar}[P_z\hat{S}_z, P_xB_y\hat{S}_z - P_xB_z\hat{S}_y - P_yB_x\hat{S}_z + P_yB_z\hat{S}_x + P_zB_x\hat{S}_y - P_zB_y\hat{S}_x]\end{aligned} \tag{3.38}$$

Here, we immediately have used the results for the commutator $[\hat{\rho},\hat{H}]$ and skipped λ in front.

17

3 The Equations of motion

After working out the commutators we have:

$$\begin{aligned}\frac{1}{\hbar}[\hat{\rho},[\hat{\rho},\hat{H}]] &= [P_x(P_yB_y + P_zB_z) - (P_y^2 + P_z^2)B_x]\hat{S}_x \\ &+ [P_y(P_xB_x + P_zB_z) - (P_x^2 + P_z^2)B_y]\hat{S}_y \\ &+ [P_z(P_xB_x + P_yB_y) - (P_x^2 + P_y^2)B_z]\hat{S}_z \,,\end{aligned} \quad (3.39)$$

or written as vector:

$$(\star) = \begin{pmatrix} P_x(P_yB_y + P_zB_z) - (P_y^2 + P_z^2)B_x \\ P_y(P_xB_x + P_zB_z) - (P_x^2 + P_z^2)B_y \\ P_z(P_xB_x + P_yB_y) - (P_x^2 + P_y^2)B_z \end{pmatrix} = \begin{pmatrix} P_x \\ P_y \\ P_z \end{pmatrix}(P_xB_x + P_yB_y + P_zB_z)$$

$$- \begin{pmatrix} B_x \\ B_y \\ B_z \end{pmatrix}(P_x^2 + P_y^2 + P_z^2) \,. \quad (3.40)$$

This can be written in shorter form as:

$$(\star) = \mathbf{P}(\mathbf{P}\cdot\mathbf{B}) - \mathbf{B}(\mathbf{P}\cdot\mathbf{P}) = \mathbf{P}\times(\mathbf{P}\times\mathbf{B}) \quad (3.41)$$

Therefore, the von Neumann equation:

$$\frac{d\hat{\rho}}{dt} = \frac{i}{\hbar}\left[\hat{\rho},\hat{H}\right] - \frac{\lambda}{\hbar}\left[\hat{\rho},\left[\hat{\rho},\hat{H}\right]\right] \quad (3.42)$$

becomes:

$$\frac{d\mathbf{P}}{dt} = \frac{1}{\hbar}(\mathbf{P}\times\mathbf{B}) - \lambda\mathbf{P}\times(\mathbf{P}\times\mathbf{B}) \,. \quad (3.43)$$

Alternatively, we can write this equation as:

$$\sum_{\alpha=1}^{3}\frac{dP_\alpha}{dt}\hat{S}_\alpha = \sum_{\alpha=1}^{3}\left[-\lambda[P_\alpha(\mathbf{P}\cdot\mathbf{B}) - B_\alpha(\mathbf{P}\cdot\mathbf{P})] + \frac{1}{\hbar}\sum_{\beta,\gamma=1}^{3}\epsilon_{\alpha\beta\gamma}P_\beta B_\gamma\right]\hat{S}_\alpha \,, \quad (3.44)$$

where $\epsilon_{\alpha\beta\gamma}$ is the Levi-Cevita tensor and $\{\alpha,\beta,\gamma\} = \{1,2,3\}$ stands for $\{x,y,z\}$.

Some words about the dimension: $\hat{\rho}$ has the dimension of $[\hbar] = Js$. $\mathbf{P} = \langle\hat{\mathbf{S}}\rangle/\hbar S$ with $|\langle\hat{\mathbf{S}}\rangle| = \hbar S$ has the dimension 1 (dimensionless). $\hat{H} = -\mathbf{B}\cdot\hat{\mathbf{S}}/\hbar$ has the dimension J, because $\mathbf{B}$ shall have this dimension, e.g. $\mathbf{B} = J(\mathbf{S}_{n-1} + \mathbf{S}_{n+1})/\hbar$, with $[J] = J$ (J on the left hand side is the exchange constant and on the right hand side the dimension Joule). $\hat{\mathbf{S}}/\hbar$ is dimensionless. λ has the dimension of $1/(Js)$. Therefore, the dimensions of both sides of the von Neumann equation (3.42) is J. In the case of Eq. (3.43) we have the dimension $1/s$ on both sides.

3.1 Derivation of the Landau-Lifshitz equation

Concerning the classical Landau-Lifshitz equation: here, both sides of the equation have the dimension $1/s$ and λ has the dimension of $1/(Ts)$. This seems to be in contradiction to the quantum mechanical λ in the von Neumann equation (3.17) resp. Heisenberg equation (3.20) which has the dimension $1/(Js)$. However, there is no problem:

$$\left[\frac{\gamma}{\mu_S}\right] = \frac{\frac{1}{Ts}}{\frac{J}{T}} = \frac{1}{Js} = \left[\frac{1}{\hbar}\right] . \tag{3.45}$$

The left hand side corresponds to the classical Landau-Lifshitz equation and the right hand side to Eq. (3.43). The difference between both equations is just the different signs of the first (precessional) term: Eq. (3.43): $+$ and in the Landau-Lifshitz equation (3.21): $-$. The difference appears due to different definitions of γ: electron: $-e$ or nucleus: $+e$.

In the case of λ we can write:

$$\lambda_{\text{qm}} = \frac{\tilde{\lambda}_{\text{qm}}}{\hbar} = \frac{\tilde{\lambda}_{\text{qm}}\gamma}{\mu_S} = \frac{\lambda_{\text{cl}}}{\mu_S} . \tag{3.46}$$

Concerning the dimensions: λ_{qm} has the dimension $1/(Js)$, $\tilde{\lambda}_{\text{qm}}$ is dimensionless, γ has the dimension $1/(Ts)$, μ_S the dimension J/T and λ_{cl} the dimension $1/(Ts)$.

The third way to derive the Landau-Lifshitz equation (3.21) shall be discussed in the following. Therefore, some words about the way how to calculate the spin dynamics using the time dependent Schrödinger equation (3.11). In this case, we have to solve first the Schrödinger equation and then to calculate the expectation value $\langle \hat{\mathbf{S}} \rangle = \langle \psi(t) | \hat{\mathbf{S}} | \psi(t) \rangle$.

Alternatively, we can look for the time development of $\mathbf{m}(t) = \langle \psi(t) | \hat{\sigma} | \psi(t) \rangle$:

$$\frac{d\mathbf{m}}{dt} = \langle \dot{\psi} | \hat{\sigma} | \psi \rangle + \langle \psi | \hat{\sigma} | \dot{\psi} \rangle . \tag{3.47}$$

In the following, the description shall be restricted to $S = 1/2$. In this case $|\psi(t)\rangle$ is given by

$$|\psi(t)\rangle = e^{-\frac{i\phi}{2}} \cos(\theta/2) |\uparrow\rangle + e^{\frac{i\phi}{2}} \sin(\theta/2) |\downarrow\rangle , \tag{3.48}$$

with $|\uparrow\rangle$ and $|\downarrow\rangle$ the normal basis vectors in the Zeeman basis. $\hat{\sigma} = (\hat{\sigma}_x, \hat{\sigma}_y, \hat{\sigma}_z)$ are the Pauli matrices[1]. With this information, it is easy to show that

$$\mathbf{m} = \begin{pmatrix} m_x \\ m_y \\ m_z \end{pmatrix} = \begin{pmatrix} \cos\phi \sin\theta \\ \sin\phi \sin\theta \\ \cos\theta \end{pmatrix} \tag{3.49}$$

---
[1] The connection between spin and Pauli matrices is given by: $\hat{\mathbf{S}} = \hbar\hat{\sigma}/2$.

3 The Equations of motion

holds. Further, $\mathbf{m}(t)$ is normalized. This is not the case for $\langle \hat{\mathbf{S}} \rangle$.

Let us come back to equation (3.47). The time development of the z-component m_z is given by:

$$\frac{dm_z}{dt} = \langle \dot{\psi} | \hat{\sigma}_z | \psi \rangle + \langle \psi | \hat{\sigma}_z | \dot{\psi} \rangle \,. \tag{3.50}$$

Furthermore, $|\dot{\psi}\rangle = \frac{d}{dt}|\psi\rangle$ and $\langle \dot{\psi}| = \frac{d}{dt}\langle \psi|$ are represented by the time dependent Schrödinger equation (3.11) and the corresponding conjugate transposed Schrödinger equation (3.13).

Now, we make the same assumption as before in the case of the von Neumann equation and assume that the Hamilton operator can be written as:

$$H = -\frac{1}{\hbar}\mathbf{B} \cdot \hat{\mathbf{S}} = -\frac{1}{2}\mathbf{B} \cdot \hat{\sigma} = -\frac{1}{2}(B_x \hat{\sigma}_x + B_y \hat{\sigma}_y + B_z \hat{\sigma}_z) \,. \tag{3.51}$$

Therefore, we get:

$$\begin{aligned} \langle H \rangle &= \langle \psi | H | \psi \rangle = -\frac{1}{2}(B_x \langle \psi | \hat{\sigma}_x | \psi \rangle + B_y \langle \psi | \hat{\sigma}_y | \psi \rangle + B_z \langle \psi | \hat{\sigma}_z | \psi \rangle) \\ &= -\frac{1}{2}(B_x m_x + B_y m_y + B_z m_z) = -\frac{1}{2}\mathbf{B} \cdot \mathbf{m} \end{aligned} \tag{3.52}$$

Inserting the Schrödinger equations (3.11) and (3.13) together with Eq. (3.51) and (3.52) in Eq. (3.50) we get:

$$\begin{aligned} \frac{dm_z}{dt} =& -\frac{i}{2\hbar}\left(B_x \langle \psi | [\hat{\sigma}_x, \hat{\sigma}_z] | \psi \rangle + B_y \langle \psi | [\hat{\sigma}_y, \hat{\sigma}_z] | \psi \rangle\right) \\ &+ \frac{\lambda}{2\hbar}\left(B_x \langle \psi | [\hat{\sigma}_x, \hat{\sigma}_z]_+ | \psi \rangle + B_y \langle \psi | [\hat{\sigma}_y, \hat{\sigma}_z]_+ | \psi \rangle + 2B_z\right) \\ &- \frac{\lambda}{\hbar}(\mathbf{B} \cdot \mathbf{m}) m_z \end{aligned} \tag{3.53}$$

Here, we have used the definition of $m_z = \langle \psi | \hat{\sigma}_z | \psi \rangle$ and $\hat{\sigma}_z \hat{\sigma}_z = \hat{1}$, where $\hat{1}$ is the unity matrix. The same is true for the other Pauli matrices: $\hat{\sigma}_x \hat{\sigma}_x = \hat{\sigma}_y \hat{\sigma}_y = \hat{1}$. In Eq. (3.53) the $[\hat{\sigma}_\alpha, \hat{\sigma}_\beta] = \hat{\sigma}_\alpha \hat{\sigma}_\beta - \hat{\sigma}_\beta \hat{\sigma}_\alpha$ are commutators while the $[\hat{\sigma}_\alpha, \hat{\sigma}_\beta]_+ = \hat{\sigma}_\alpha \hat{\sigma}_\beta + \hat{\sigma}_\beta \hat{\sigma}_\alpha$ are anticommutators. The commutators are given by $[\hat{\sigma}_\alpha, \hat{\sigma}_\beta] = 2i\epsilon_{\alpha,\beta,\gamma}\hat{\sigma}_\gamma$. This is independent of S, only the Pauli matrices change with changing S. The anticommutators $[\hat{S}_\alpha, \hat{S}_\beta]_+$ change with S. In the case of $S = 1/2$ the anticommutators are given by: $[\hat{\sigma}_\alpha, \hat{\sigma}_\beta]_+ = 2\delta_{\alpha,\beta}\hat{1}$. However, this is not the case for $S > 1/2$.

Working out the anticommutators and commutators we get:

$$\frac{dm_z}{dt} = \frac{1}{\hbar}[m_x B_y - m_y B_x] - \frac{\lambda}{\hbar}[m_z(\mathbf{B} \cdot \mathbf{m}) - B_z \underbrace{(\mathbf{m} \cdot \mathbf{m})}_{=1}] \tag{3.54}$$

3.1 Derivation of the Landau-Lifshitz equation

Here, we have used the definitions for $m_x = \langle\psi|\hat{\sigma}_x|\psi\rangle$ and $m_y = \langle\psi|\hat{\sigma}_y|\psi\rangle$ and the fact that **m** is normalized: $\mathbf{m}^2 = 1$.

The last equation can be written in a more compact form as:

$$\frac{dm_z}{dt} = \frac{1}{\hbar}[\mathbf{m}\times\mathbf{B}]_z - \frac{\lambda}{\hbar}[\mathbf{m}\times(\mathbf{m}\times\mathbf{B})]_z \qquad (3.55)$$

The equations for m_x and m_y can be derived in a similar way. Therfore, we will find at the end that:

$$\frac{d\mathbf{m}}{dt} = \frac{1}{\hbar}\mathbf{m}\times\mathbf{B} - \frac{\lambda}{\hbar}\mathbf{m}\times(\mathbf{m}\times\mathbf{B}) \qquad (3.56)$$

holds. This equation is similar to Eq. (3.43). **m** and **P** base on the spin expectation values $\langle\hat{\mathbf{S}}\rangle$ and are normalized. However, there is a small detail: in Eq. (3.43) there is no $1/\hbar$ in front of the damping term and λ has the dimension $1/JS$. Here, in this equation λ is dimensionless, however, due to the $1/\hbar$ we get the same dimension on both sides as in Eq. (3.43): $1/s$ (**B** has the dimension J). This means we can write this equation also as:

$$\frac{d\mathbf{m}}{dt} = \frac{1}{\hbar}\mathbf{m}\times\mathbf{B} - \lambda'\mathbf{m}\times(\mathbf{m}\times\mathbf{B})\,, \qquad (3.57)$$

where $\lambda' = \lambda/\hbar$ is equal to the λ in Eq. (3.43).

Now, the question arises: Where have we changed the dimension of λ, because we have derived Eq. (3.43) from the von Neuman equation (3.17) which we have got from the time dependent Schrödinger equation (3.11)? The answer can be found in Eq. (3.12):

$$i\hbar\frac{d}{dt}|\psi\rangle = \left(H - i\lambda\Big[H,|\psi\rangle\langle\psi|\Big]\right)|\psi\rangle\,. \qquad (3.58)$$

If we assume that $|\psi\rangle$ is dimensionless, $|\psi\rangle\langle\psi|$ is dimensionless too. The same is true for λ. This was the case in the last consideration. Therefore, we need the additional $1/\hbar$ to get the correct dimensuion. However, if we set $|\psi\rangle\langle\psi| = \rho$ equal to Eq. (3.32) we change the dimension of $|\psi\rangle\langle\psi|$ to Js and compensate this with the dimension $1/Js$ for λ. This was the case in Eq. (3.43). In this case the additional $1/\hbar$ does not appear because the equation has already the correct dimension. In other words, the $1/\hbar$ is already included in λ.

As discussed before, the anticommutator relation: $[\hat{\sigma}_\alpha,\hat{\sigma}_\beta]_+ = 2\delta_{\alpha,\beta}\hat{1}$ only holds for $S = 1/2$ which means that σ_α are the real Pauli matrices. In these cases the terms with the anticommutators in Eq. (3.53) become zero. However, the simulations show a good agreement between the Landau-Lifshitz equation and the time development of the expectation value $\langle\hat{\mathbf{S}}\rangle$ also for $S > 1/2$ if the the Hamiltonian can be written as assumed in Eq. (3.33) (linear in $\hat{\mathbf{S}}$). This means

3 The Equations of motion

Eq. (3.56) also holds in these cases. This has also be shown due to Eq. (3.43) which is valid for all S. This also means that the terms with the anticommutators become zero. However, in the cases $S > 1/2$ not because of the anticommutators themselves.

So far, we have only discussed the cases where we have assumed that $\hat{H}$ can be written as $\hat{H} = -\mathbf{B} \cdot \hat{\mathbf{S}}_n/\hbar$, with the assumption that $\mathbf{B}$ is independent of $\hat{\mathbf{S}}_n$: $[\hat{\mathbf{S}}_n, \mathbf{B}] = 0$. This is the case for nearly all the Heisenberg model Hamilton operators e.g.:
(isotropic) exchange interaction

$$\hat{\mathcal{H}}_J = -\frac{J}{\hbar^2} \sum_{\langle nm \rangle} \hat{\mathbf{S}}_n \cdot \hat{\mathbf{S}}_m , \qquad (3.59)$$

anisotropic exchange interaction

$$\hat{\mathcal{H}}_{\text{XYZ}} = -\frac{1}{\hbar^2} \sum_{\langle nm \rangle} \left(J_x \hat{S}_n^x \hat{S}_m^x + J_y \hat{S}_n^y \hat{S}_m^y + J_z \hat{S}_n^z \hat{S}_m^z \right) , \qquad (3.60)$$

dipole-dipole interaction

$$\hat{\mathcal{H}}_{\text{DD}} = -\frac{\omega}{\hbar^2} \sum_{n<m} \frac{3(\hat{\mathbf{S}}_n \cdot \mathbf{e}_{nm})(\hat{\mathbf{S}}_m \cdot \mathbf{e}_{nm}) - \hat{\mathbf{S}}_n \cdot \hat{\mathbf{S}}_m}{r_{nm}^3} , \qquad (3.61)$$

Dzyaloshinsky-Moriya interaction

$$\hat{\mathcal{H}}_{\text{DM}} = -\frac{1}{\hbar^2} \mathbf{D}_{\text{DM}} \cdot \sum_{\langle nm \rangle} \hat{\mathbf{S}}_n \times \hat{\mathbf{S}}_m , \qquad (3.62)$$

coupling to an external field (Zeeman term)

$$\hat{\mathcal{H}}_B = -\frac{g\mu_B}{\hbar} \sum_n \mathbf{B} \cdot \hat{\mathbf{S}}_n . \qquad (3.63)$$

In all these cases the commutator $[\hat{\mathbf{S}}_n, \hat{\mathcal{H}}]$ is given by:

$$[\hat{\mathbf{S}}_n, \hat{\mathcal{H}}] = \hat{\mathbf{S}}_n \times \frac{\partial \hat{\mathcal{H}}}{\partial \hat{\mathbf{S}}_n} \qquad (3.64)$$

However, there are also Hamilton operators which do not fit into this scheme, Hamilton operators which are not "linear" in $\hat{\mathbf{S}}_n$ e.g.:
uniaxial anisotropy (here easy axis in z-direction)

$$\hat{\mathcal{H}}_{D_z} = -\frac{D_z}{\hbar^2} \sum_n (\hat{S}_n^z)^2 , \qquad (3.65)$$

3.1 Derivation of the Landau-Lifshitz equation

or the biquadratic exchange

$$\hat{\mathcal{H}}_{\text{bi}} = -\frac{J_{\text{bi}}}{\hbar^2} \sum_{\langle nm \rangle} \left(\hat{\mathbf{S}}_n \cdot \hat{\mathbf{S}}_m\right)^2 \tag{3.66}$$

In these cases we get additional corrections of the order $\hbar$

$$[\hat{\mathbf{S}}_n, \hat{\mathcal{H}}] = \hat{\mathbf{S}}_n \times \frac{\partial \hat{\mathcal{H}}}{\partial \hat{\mathbf{S}}_n} + \mathcal{O}(\hbar) \ . \tag{3.67}$$

These corrections describe the additional quantum effects. In the classical limit $S \to \infty$ and $\hbar \to 0$ these additional terms vanish and we reproduce the classical result Eq. (3.19) again.

What are the consequences? If we compare the trajectory of the quantum mechanical expectation value $\langle \hat{\mathbf{S}}_n \rangle$ with the corresponding trajectory of the classical spin $\mathbf{S}_n$ we will notice that there is a perfect agreement if the Hamilton operator is linear in $\hat{\mathbf{S}}_n$. However, we will notice a deviation if the Hamilton operator contains quadratic or higher order terms in $\hat{\mathbf{S}}_n$. This also means that the Ehrenfest theorem, which says that the quantum mechanical expectation values behave classical, is generally not correct. It only works if the the Heisenberg Hamilton operators are linear. To clarify this point we can use the following correspondance:

$$\mathbf{F}(\mathbf{r}) = -\nabla_\mathbf{r} U(\mathbf{r}) \quad \leftrightarrow \quad \mathbf{H}_{\text{eff}} = -\nabla_{\hat{\mathbf{S}}} \hat{\mathcal{H}}(\hat{\mathbf{S}}) \ . \tag{3.68}$$

The effective field $\mathbf{H}_{\text{eff}}$ in the spin dynamics plays the same role as the force $\mathbf{F}(\mathbf{r})$ in classical physics. Furthermore, we can associate the Heisenberg Hamiltonian $\hat{\mathcal{H}}$ with the potential energy $U(\mathbf{r})$.

In his original work Ehrenfest has assumed that $\langle \mathbf{F}(\mathbf{r}) \rangle = -\langle \nabla U(\mathbf{r}) \rangle$ can be replaced by $\mathbf{F}(\langle \mathbf{r} \rangle) = -\nabla U(\langle \mathbf{r} \rangle)$. However, $\langle \mathbf{F}(\mathbf{r}) \rangle = \mathbf{F}(\langle \mathbf{r} \rangle)$ only holds if $\mathbf{F}$ is linear in $\mathbf{r}$.

Moreover, the Ehrenfest theorem ignores another important quantum effect the entanglement [14, 15]. Sofar, we have assumed $|\langle \hat{\mathbf{S}} \rangle| = \hbar S$, however this is not the general scenario. The entanglement lead to $|\langle \hat{\mathbf{S}} \rangle| \leq \hbar S$. The best example is the antiferromagnetic groundstate e.q. for two spins with $S = 1/2$:

$$|\psi\rangle = \frac{1}{2}\left(|\uparrow\downarrow\rangle - |\downarrow\uparrow\rangle\right) \ . \tag{3.69}$$

In this case we find zero expectation values $\langle \hat{\mathbf{S}}_n \rangle = 0 \Leftrightarrow |\langle \hat{\mathbf{S}}_n \rangle| = 0$ for both spins $n \in \{1, 2\}$.

Furthermore, $|\langle \hat{\mathbf{S}}_n \rangle|$ is not necessarily constant during the dynamics. As the expectation values $\langle \hat{\mathbf{S}}_n \rangle = \langle \psi | \hat{\mathbf{S}}_n | \psi \rangle$, $|\langle \hat{\mathbf{S}}_n \rangle| = |\langle \psi | \hat{\mathbf{S}}_n | \psi \rangle|$ strongly depends on the

3 The Equations of motion

wave function $|\psi\rangle$, which is not necessarily constant in time. This means that in general we expect a total different dynamics of quantum mechanical spins with respect to classical ones. Only in the limit $S \to \infty$ and $\hbar \to 0$ we expect the same behavior. However, there are three exceptional cases: First: single spins. Here, we have no entanglement and therefore a constant absolut value $|\langle \hat{\mathbf{S}}_n \rangle| = \hbar S$. In this case we can assume a classical behavior of the expectation value $\langle \hat{\mathbf{S}} \rangle$ as long as the Hamiltonian can be written as a Zeeman term $\hat{\mathcal{H}} = -\mathbf{B} \cdot \hat{\mathbf{S}}/\hbar$. Second: coherent states [16, 17, 18]. These, states are constructed in such a way that the expectation values correspond to the classical results. And the third exceptional case are linear excitations of the ferromagnetic ground state. In this case the wave function $|\psi\rangle$ is approximately the ferromagnetic ground state: $|\psi\rangle \approx |\text{FM}\rangle$. This means a single state and not an entanglement. Therefore, we can assume $|\langle \hat{\mathbf{S}}_n \rangle| \approx \hbar S$ for all times and a classical behavior as long as we can write the Hamiltonian as effective Zeeman term $\hat{\mathcal{H}} = -\mathbf{B} \cdot \hat{\mathbf{S}}_n/\hbar$ (see discussion before).

Until now, we have discussed the Landau-Lifshitz equation for a magnetic moment Eq. (3.21) and the way how to derive this equation. This equation can be used to describe the dynamics of a classical spin on the atomic level. On the microscopic level we have to replace the magnetic moment $\mathbf{S} = \mu/\mu_S$ by the magnetization $\mathbf{M} = \mathbf{S} M_S$.

Then we can take the Landau-Lifshitz equation:

$$\frac{d\mathbf{S}}{dt} = -\frac{\gamma}{\mu_S}\mathbf{S} \times \mathbf{H}_{\text{eff}} - \frac{\lambda}{\mu_S}\mathbf{S} \times (\mathbf{S} \times \mathbf{H}_{\text{eff}}) \,. \tag{3.70}$$

and define $\mathbf{H} = \mathbf{H}_{\text{eff}}/\mu_S$. $\mathbf{H}_{\text{eff}}$ has the dimension J, μ_S the dimension J/T. Therefore, $\mathbf{H}$ is a real field with the dimension T and the Landau-Lifshitz equation becomes:

$$\frac{d\mathbf{S}}{dt} = -\gamma \mathbf{S} \times \mathbf{H} - \lambda \mathbf{S} \times (\mathbf{S} \times \mathbf{H}) \,. \tag{3.71}$$

In a second step we multiply M_s (magnetization) to both sides of the Landau-Lifshitz equation. Therefore, the dimensionless $\mathbf{S}$ becomes $\mathbf{M} = M_S \mathbf{S}$ with the dimension of the magnetization: $[M_S] = A/m$. We get:

$$\frac{d\mathbf{M}}{dt} = -\gamma \mathbf{S} \times \mathbf{H} - \lambda \mathbf{S} \times (\mathbf{M} \times \mathbf{H}) \,. \tag{3.72}$$

To make it consistent we write: $\mathbf{S} = (M_S/M_S)\mathbf{S} = \mathbf{M}/M_S$ and get finally the Landau-Lifshitz equation for the magnetization $\mathbf{M}$:

$$\frac{d\mathbf{M}}{dt} = -\gamma \mathbf{S} \times \mathbf{H} - \frac{\lambda}{M_S}\mathbf{M} \times (\mathbf{M} \times \mathbf{H}) \,. \tag{3.73}$$

3.2 The Gilbert respectively Landau-Lifshitz-Gilbert equation

At this point the questions appear: How good are the assumptions we have made so far? And, does the Landau-Lifshitz equation (3.21) describes the physics correctly? In the previous section we have seen that classical trajectories and quantum mechanical expectation values could be identical. However, we have derived the Landau-Lifshitz equation from a damped time dependent Schrödinger equation where the damping came from a non-Hermitian Hamiltonian $\hat{\mathcal{H}}$. This means the energy dissipation is derived from a phenomenological ansatz. The same is true for the original derivation of Landau-Lifshitz equation by L. D. Landau and J. M. Lifshitz in 1935. To proof the accuracy let us investigate the following simple scenario [19]. We consider a magnetic nanoparticle which is single domain. The Hamilton function is given by:

$$\mathcal{H} = H_z \hat{\mathbf{z}} \tag{3.74}$$

which gives the following set of differential (Landau-Lifshitz) equations for the magnetization components:

$$\frac{dM_x}{dt} = \gamma M_y H_z - \frac{\lambda}{M_S} M_x M_z H_z , \tag{3.75}$$

$$\frac{dM_y}{dt} = -\gamma M_x H_z - \frac{\lambda}{M_S} M_y M_z H_z , \tag{3.76}$$

$$\frac{dM_z}{dt} = \frac{\lambda}{M_S} \left(M_S^2 - M_z^2 \right) H_z , \tag{3.77}$$

Here, we are interested in the reversal time $\tau = t_f - t_i$ for a magnetization reversal in the external field H starting at $t = t_i = 0$ with a magnetization $M_z^i \approx -M_S$. At the end $(t = t_f)$ the magnetization is given by $M_z^f \approx +M_S$. The reversal time can be calculated by solving the differential equation for M_z. The other two differential equations can be skipped because we are not interested in the reversal process itself.

The differential equation (3.77) can be simply solved by separation of the variables. After a simple algebra we get the result for the reversal time:

$$\tau = \frac{1}{\lambda M_S^2 H_z} \ln \left(\frac{(M_S + M_z^f)(M_S - M_z^i)}{(M_S - M_z^f)(M_S + M_z^i)} \right) . \tag{3.78}$$

The result of this calculation is a reversal time which is indirect proportional to the damping: $\tau \propto 1/\lambda$. This means that the reversal time τ decreases with increasing damping. Or, in other words, the reversal becomes faster with higher damping. This is not realistic. A higher friction should lead to a slower motion.

3 The Equations of motion

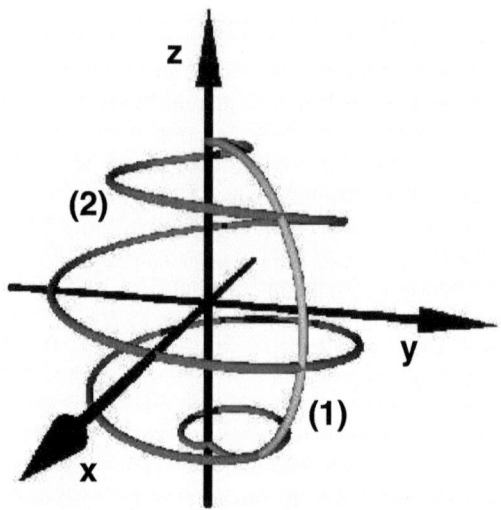

Figure 3.1: Possible reversal paths of the nanoparticle: (1) direct reversal (large damping) and (2) precessional motion (small damping).

Therefore, we can say that the Landau-Lifshitz equation does not describe the physics correctly.

Fig. 3.1 shows the possible reversal paths of the nanoparticle. We expect a precessional motion if the damping is small. In this case the precessional term $-\gamma \mathbf{M} \times \mathbf{H}$ of the Landau-Lifshitz equation is dominating. In the limit of a large damping the relaxation term $\lambda/M_S \mathbf{M} \times (\times \mathbf{H})$ becomes dominating and we expect a damped direct reversal. With larger damping the friction becomes larger and the reversal becomes more direct. Therefore, the reversal time becomes smaller ($\lambda \leq 1$) and the reversal faster. However, in the case of a huge damping $\lambda \gg 1$ we have a direct reversal which becomes slower with increasing damping. However, this is not the case for the Landau-Lifshitz equation.

In 1956 T. L. Gilbert proposed to replace the phenomenological damping term $-\lambda \mathbf{M} \times (\mathbf{M} \times \mathbf{H})$ by $-\alpha \mathbf{M} \times \partial \mathbf{M}/\partial t$ [20, 21]. Therefore, the new equation can be written as:

$$\frac{d\mathbf{M}}{dt} = -\gamma \mathbf{M} \times \mathbf{H} - \frac{\alpha}{M_S} \mathbf{M} \times \frac{d\mathbf{M}}{dt}. \qquad (3.79)$$

3.2 The Gilbert respectively Landau-Lifshitz-Gilbert equation

and in the case of magnetic moments:

$$\frac{d\mathbf{S}}{dt} = -\frac{\gamma}{\mu_S}\mathbf{S} \times \mathbf{H}_{\text{eff}} - \alpha \mathbf{S} \times \frac{d\mathbf{S}}{dt} . \tag{3.80}$$

This is the so called Gilbert equation. The transformation from the Gilbert equation for the magnetization $\mathbf{M}$ through the Gilbert equation for the normalized magnetic moment $\mathbf{S}$ is the same as before for the Landau-Lifshitz equation.

In the following it shall be demonstrated that the Gilbert equation shows the correct behavior with respect to the damping. The similar calculation as before, however with the Gilbert equation instead of the Landau-Lifshitz equation, leads to the following reversal time of the nanoparticle:

$$\tau_G = \frac{1+\alpha^2}{\alpha\gamma M_S^2 H_z} \ln\left(\frac{(M_S + M_z^f)(M_S - M_z^i)}{(M_S - M_z^f)(M_S + M_z^i)}\right) . \tag{3.81}$$

In the case of the Gilbert equation the reversal time τ is proportional to $\left(1+\alpha^2\right)/\alpha$ and in the limit of a large damping $\alpha \gg 1$, τ is directly proportional to α. This means the reversal becomes slower with increasing friction in this limit.

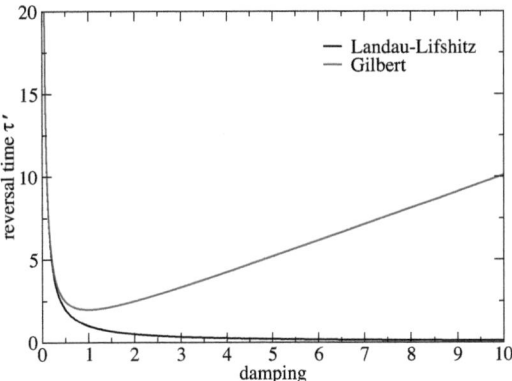

Figure 3.2: Reversal time $\tau' = \tau[\gamma M_S^2 H/\ln(\ldots)]$, of a single domain nanoparticle calculated by the Landau-Lifshitz and Gilbert equation as function of the damping: λ/γ in the case of the Landau-Lifshitz equation and α in the case of the Gilbert damping.

27

3 The Equations of motion

Fig. 3.2 shows the reversal times calculated with the Landau-Lifshitz equation τ and the Gilbert equation τ_G. It is interesting to see that in the limit of a small damping both equations show the same reversal time. This means that in this limit both equations are identical. This can be seen also when we write the implicit Gilbert equation (3.80) in its explicit form. To do so we have to multiply both sides of the differential equation with $\mathbf{S}\times$ from the left. This leads to:

$$\begin{aligned}\mathbf{S} \times \frac{d\mathbf{S}}{dt} &= -\frac{\gamma}{\mu_S}\mathbf{S} \times (\mathbf{S} \times \mathbf{H}_{\text{eff}}) + \alpha \mathbf{S} \times \left(\mathbf{S} \times \frac{d\mathbf{S}}{dt}\right) \\ &= -\frac{\gamma}{\mu_S}\mathbf{S} \times (\mathbf{S} \times \mathbf{H}_{\text{eff}}) + \alpha \mathbf{S}\left(\mathbf{S} \cdot \frac{d\mathbf{S}}{dt}\right) - \alpha \frac{d\mathbf{S}}{dt} \ . \end{aligned} \quad (3.82)$$

In the second step we have used:

$$\mathbf{a} \times (\mathbf{b} \times \mathbf{c}) = \mathbf{b}(\mathbf{a} \cdot \mathbf{c}) - \mathbf{c}(\mathbf{a} \cdot \mathbf{b}) \quad (3.83)$$

and $\mathbf{S}^2 = 1$. The second term can be skipped because $\mathbf{S} \cdot \partial \mathbf{S}/\partial t$ is equal to zero. The reason is that the spin $\mathbf{S}$ and the vector which describes the development in time $\partial \mathbf{S}/\partial t$ are perpendicular to each other: $\mathbf{S} \perp \partial \mathbf{S}/\partial t$, and therefore the scalar product $\mathbf{S} \cdot \partial \mathbf{S}/\partial t = 0$. This means:

$$\mathbf{S} \times \frac{d\mathbf{S}}{dt} = -\frac{\gamma}{\mu_S}\mathbf{S} \times (\mathbf{S} \times \mathbf{H}_{\text{eff}}) - \alpha \frac{d\mathbf{S}}{dt} \ . \quad (3.84)$$

Inserting this into the Gilbert equation (3.80) leads to the following equation:

$$\frac{d\mathbf{S}}{dt} = -\frac{\gamma}{(1+\alpha^2)\mu_S}\mathbf{S} \times \mathbf{H}_{\text{eff}} - \frac{\gamma\alpha}{(1+\alpha^2)\mu_S}\mathbf{S} \times (\mathbf{S} \times \mathbf{H}_{\text{eff}}) \ , \quad (3.85)$$

with $\mathbf{H}_{\text{eff}} = -\frac{\partial \mathcal{H}}{\partial \mathbf{S}}$. This equation is called Landau-Lifshitz-Gilbert (LLG) equation. The corresponding equation for the magnetization is given by:

$$\frac{d\mathbf{M}}{dt} = -\frac{\gamma}{(1+\alpha^2)}\mathbf{M} \times \mathbf{H} - \frac{\gamma\alpha}{(1+\alpha^2)M_S}\mathbf{M} \times (\mathbf{M} \times \mathbf{H}) \ , \quad (3.86)$$

The difference between this equation and the Landau-Lifshitz equation (3.21) resp. (3.73) is that now both terms on the right hand side are damped. This means that precessional motion as well as the relaxation is damped. This was not the case in the Landau-Lifshitz equation. In this case we have an undamped precession and a damping induced relaxation. However, it makes more sense to speak about two possible motions (precession and relaxation) which are subject to the damping process.

In the limit of small damping $\alpha \to 0$ the $1-\alpha^2$ becomes ≈ 1 and the Landau-Lifshitz-Gilbert equation (3.85) becomes identical to the Landau-Lifshitz equation

3.2 The Gilbert respectively Landau-Lifshitz-Gilbert equation

(3.21) with $\alpha = \lambda$. This explains the agreement of the reversal times in Fig. 3.2 in the limit of a small damping.

In real systems the damping constants α are small. Furthermore, the observed spin dynamics in experiments can be reproduced by solving the Landau-Lifshitz-Gilbert equation. This means: first, the Landau-Lifshitz-Gilbert equation gives an adequate description and, second, the same is true for the Landau-Lifshitz equation.

The Gilbert equation in spherical coordinates

In the last subsections we have investigated the Gilbert, the Landau-Lifshitz, and the Landau-Lifshitz-Gilbert equation as well as the corresponding quantum mechanical time dependent Schrödinger resp. Heisenberg equation. All these equations have been investigated in the framework of cartesian coordinates.

This subsection shall show how to get the Gilbert equation in spherical coordinates. The transformation to spherical coordinates becomes important for the analytical calculation. The first reason is the reduction of the number of coupled differential equations. The Gilbert equation in cartesian coordinates corresponds to three coupled differential equations one for each direction S_x, S_y and S_z. In the case of spherical coordinates the number of differential equations reduces to two due to the assumption that $S = \mathrm{const.} = 1$. The differential equations in this case are one for the angle θ and one for angle ϕ. The second reason comes from the symmetry. Due to the constant length of the spin the motion can be described as a point on the surface of a sphere. Therefore, spherical coordinates seem to be most adequate. However, for computer simulations it is better to choose cartesian coordinates. Spherical coordinates are less adequate in this case because of the higher numerical effort in the calculation of the trigonometric functions which makes the calculation slower and less accurate. Nevertheless, let us come back to the derivation of the Gilbert equation in spherical coordinates which shall be described in the following.

The Gilbert equation in cartesian coordinates is given as:

$$\dot{\mathbf{S}}_n = \frac{\gamma}{\mu_S} \mathbf{S}_n \times \mathbf{H}_{eff}^n + \alpha \mathbf{S}_n \times \dot{\mathbf{S}}_n \, , \tag{3.87}$$

with $\gamma > 0$ and $\mathbf{H}_{eff}^n = -\partial \mathcal{H}/\partial \mathbf{S}_n$.

The index n numbers the spins and corresponds to a certain position in the lattice. Such a numeration is necessary if we deal with more than one spin.

The spherical coordinates are given by:

$$\mathbf{S}_n = S \mathbf{e}_S = \begin{pmatrix} S \sin \theta_n \cos \phi_n \\ S \sin \theta_n \sin \phi_n \\ S \cos \theta_n \end{pmatrix} , \tag{3.88}$$

3 The Equations of motion

where

$$\mathbf{e}_S = \begin{pmatrix} \sin\theta_n \cos\phi_n \\ \sin\theta_n \sin\phi_n \\ \cos\theta_n \end{pmatrix} \quad \mathbf{e}_\theta = \begin{pmatrix} \cos\theta_n \cos\phi_n \\ \cos\theta_n \sin\phi_n \\ -\sin\theta_n \end{pmatrix} \quad \mathbf{e}_\phi = \begin{pmatrix} -\sin\phi_n \\ \cos\phi_n \\ 0 \end{pmatrix} \quad (3.89)$$

are the normalized basis vectors with $\mathbf{e}_\phi \times \mathbf{e}_S = \mathbf{e}_\theta$, respectively $\mathbf{e}_S \times \mathbf{e}_\phi = -\mathbf{e}_\theta$ and cyclic commutations.

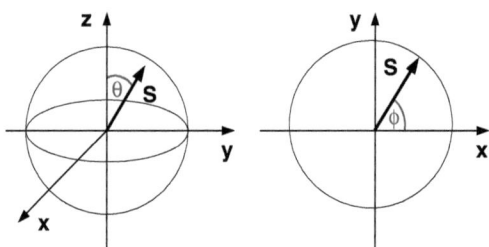

Figure 3.3: Spherical coordinates

Therefore, we can write for the spin:

$$\dot{\mathbf{S}}_n = \dot{S}\mathbf{e}_S + S\dot{\mathbf{e}}_S = \dot{S}\mathbf{e}_S + S\dot{\theta}_n\mathbf{e}_\theta + S\sin\theta_n\dot{\phi}_n\mathbf{e}_\phi \;, \quad (3.90)$$

as well as

$$\frac{\partial \mathcal{H}}{\partial \mathbf{S}_n} = \frac{\partial \mathcal{H}}{\partial S}\mathbf{e}_S + \frac{1}{S}\frac{\partial \mathcal{H}}{\partial \theta_n}\mathbf{e}_\theta + \frac{1}{S\sin\theta_n}\frac{\partial \mathcal{H}}{\partial \phi_n}\mathbf{e}_\phi \;. \quad (3.91)$$

If we assume that $S = 1 = \text{const.}$, we get:

$$\mathbf{S}_n = \mathbf{e}_S \;, \quad (3.92)$$

$$\dot{\mathbf{S}}_n = \dot{\theta}_n\mathbf{e}_\theta + \sin\theta_n\dot{\phi}_n\mathbf{e}_\phi \;, \quad (3.93)$$

and

$$\frac{\partial \mathcal{H}}{\partial \mathbf{S}_n} = \frac{\partial \mathcal{H}}{\partial \theta_n}\mathbf{e}_\theta + \frac{1}{\sin\theta_n}\frac{\partial \mathcal{H}}{\partial \phi_n}\mathbf{e}_\phi \;. \quad (3.94)$$

3.2 The Gilbert respectively Landau-Lifshitz-Gilbert equation

Then, with aid of Eq. (3.92) to (3.94) we can write:

$$\dot{\mathbf{S}}_n = \dot{\theta}_n \mathbf{e}_\theta + \sin\theta_n \dot{\phi}_n \mathbf{e}_\phi , \tag{3.95}$$

$$-\frac{\gamma}{\mu_S}\mathbf{S}_n \times \frac{\partial \mathcal{H}}{\partial \mathbf{S}_n} = -\frac{\gamma}{\mu_S}\frac{\partial \mathcal{H}}{\partial \theta_n}\mathbf{e}_\phi + \frac{\gamma}{\mu_S \sin\theta_n}\frac{\partial \mathcal{H}}{\partial \phi_n}\mathbf{e}_\theta , \tag{3.96}$$

and

$$\alpha \mathbf{S}_n \times \dot{\mathbf{S}}_n = \alpha\left(\dot{\theta}_n \mathbf{e}_\phi - \sin\theta_n \dot{\phi}_n \mathbf{e}_\theta\right) . \tag{3.97}$$

Therefore, the Gilbert Eq. (3.87) can be written in spherical coordinates as:

$$\dot{\theta}_n = -\alpha \sin\theta_n \dot{\phi}_n + \frac{\gamma}{\mu_S \sin\theta_n}\frac{\partial \mathcal{H}}{\partial \phi_n} \tag{3.98}$$

and

$$\sin\theta_n \dot{\phi}_n = \alpha \dot{\theta}_n - \frac{\gamma}{\mu_S}\frac{\partial \mathcal{H}}{\partial \theta_n} . \tag{3.99}$$

In the micromagnetic limit $\partial \mathcal{H}/\partial \phi_n$ resp. $\partial \mathcal{H}/\partial \theta_n$ have to be replaced by the variations: $\delta\mathcal{E}/\delta\phi$ resp. $\delta\mathcal{E}/\delta\theta$, with the energy density $\mathcal{E}$.

Quantum mechanical derivation of the Landau-Lifshitz equation

The described way to derive the Landau-Lifshitz-Gilbert equation is classical one and can be found in some text books. In the meantime D. Altwein has pointed out that not only the Landau-Lifshitz equation but also the Landau-Lifshitz-Gilbert equation can be derived from quantum mechanics [22].

D. Altwein has noticed that the underlying physics behind the non-Hermitian Hamiltonian $\hat{\mathcal{H}}$ in the derivation of the Landau-Lifshitz equation (Sec. 3.1) is identical to a second-order time-dependent perturbation using the Wigner-Weisskopf formalism [23]. This means, that we can understand the Hamiltonian $\hat{\mathcal{H}} = \hat{H} - i\lambda\hat{H}$ either as a complex number (matrix) with a real and an imaginary part as before in the derivation of the Landau-Lifshitz equation or we can describe $\hat{\mathcal{H}}$ by a perturbation series. In this case $\hat{\mathcal{H}} = \hat{H} - i\lambda\hat{H}$ is a perturbation of the first order in $\hat{H}$. If we represent $\hat{\mathcal{H}}$ by a perturbation series we can take into account all higher order terms:

$$\hat{\mathcal{H}} = \lim_{N\to\infty}\sum_{n=0}^{N}(-i)^n\lambda^n\hat{H} = \hat{H} - i\lambda\hat{H} - \lambda^2\hat{H} + i\lambda^3\hat{H} + \lambda^4\hat{H} - \ldots + (-i)^N\lambda^N\hat{H} . \tag{3.100}$$

3 The Equations of motion

A closer look shows that Eq. (3.100) is nothing than as a geometric series times $\hat{H} - i\lambda\hat{H}$:

$$\hat{\mathcal{H}} = (1 - \lambda^2 + \lambda^4 - \ldots - \lambda^{N-2} + \lambda^N)(\hat{H} - i\lambda\hat{H}) = \frac{\hat{H} - i\lambda\hat{H}}{1 + \lambda^2} . \quad (3.101)$$

This means, we end up with a non-Hermitian Hamiltonian similar to the one we have used to derive the Landau-Lifshitz equation, however modified by the factor $1/(1 + \lambda^2)$. This means we can immediately write down the corresponding time dependent Schrödinger equation by taking time dependent Schrödinger equation corresponding to the Landau-Lifshitz equation, Eq. (3.11), and replacing $\hbar$ by $\hbar(1 + \lambda^2)$:

$$i\hbar(1 + \lambda^2)\frac{d}{dt}|\psi(t)\rangle = (\hat{H} - i\lambda[\hat{H} - \langle\hat{H}\rangle])|\psi(t)\rangle . \quad (3.102)$$

Then we can follow the same description as before and get as the final result:

$$\frac{d\mathbf{P}}{dt} = \frac{1}{(1+\lambda^2)\hbar}(\mathbf{P} \times \mathbf{B}) - \frac{\lambda}{(1+\lambda^2)\hbar}\mathbf{P} \times (\mathbf{P} \times \mathbf{B}) . \quad (3.103)$$

After replacing $\hbar$ by γ/μ_S [see Eq. (3.45)] and after renaming λ by α, $\mathbf{B}$ by $\mathbf{H}_{\text{eff}}$ and $\mathbf{P}$ by $\mathbf{S}$, we get the Landau-Lifshitz-Gilbert equation (3.85):

$$\frac{d\mathbf{S}}{dt} = \frac{\gamma}{(1+\alpha^2)\mu_S}(\mathbf{S} \times \mathbf{B}) - \frac{\alpha\gamma}{(1+\alpha^2)\mu_S}\mathbf{S} \times (\mathbf{S} \times \mathbf{B}) . \quad (3.104)$$

Here, the damping constant $\alpha = \lambda$ is dimensionless. Therefore, we had to introduce for dimensional reasons an additional $\hbar$ [See also Eq. (3.46)].

Eq. (3.103) shows that it is not only possible to derive the Landau-Lifshitz-Gilbert equation from quantum mechanics, Eq. (3.103) also shows that there is a stronger connection between the Landau-Lifshitz equation and the Gilbert equation as we could expect from the classical description.

However, there is still the problem that the length of the expectation values $|\langle\hat{\mathbf{S}}_n\rangle|$, and therefore the absolute values of the polarization $|\mathbf{P}_n| = |\langle\hat{\mathbf{S}}_n\rangle|/\hbar S$ are not constant during the dynamics. For more details, please see the discussion in the previous section 3.1.

3.3 The Landau-Lifshitz-Gilbert-Slonczewski equation

Beside external or internal effective fields the magnetization can be influenced also by a spin-polarized current.

The motivation in using a spin-polarized electric current can be seen in the fact that we need additional devices like coils to produce a magnetic field. For the industry it is important that at the end the magnetic device is compact. It is clear that if we can skip the coils which are needed to produce the magnetic fields the magnetic device will become more compact. Furthermore, we generate magnetic fields using an electric current. However, this also means that we have energy losses during this process. It would be more effective if we could use the electric current directly to manipulate the domain wall. In these cases the magnetic devices would become more compact and more energy-efficient. Another advantage which shall be noticed here is the fact that an electric current can be constricted to very small areas, see e.g. scanning tunneling microscope experiments [24] or the ballistic transport through an atomic chain [25]. To the best of my knowledge such a constriction is not possible in the case of external fields.

The influence of a spin-polarized contains two parts: the adiabatic and the non-adiabatic spin torque [26]. The adiabatic spin torque represents the torque on the magnetization which occurs due to the conservation of the angular momentum. The electrons are assumed to follow the magnetization adiabatically. The rotation of the electron spin leads to an inverse rotation of the magnetization.

Microscopically the adiabatic spin torque corresponds to a coupling between the localized magnetic moments $\hat{\mathbf{S}}$ and the itinerant spins $\hat{\sigma}$ (sd-Hamiltonian):

$$\hat{\mathcal{H}}_{\text{sd}} = -\frac{J}{\hbar}\hat{\sigma} \cdot \hat{\mathbf{S}} \tag{3.105}$$

The non-adiabatic spin torque describes effects which appear due to electrons which do not follow the magnetization adiabatically. The microscopic description here are scattering processes of the electrons.

Taking into account both processes we can write down the following continuity equation for the itinerant spins:

$$\frac{d\hat{\sigma}}{dt} + \nabla \hat{J} = \frac{i}{\hbar}[\,\hat{\mathcal{H}}_{\text{sd}}, \hat{\sigma}] + \frac{i}{\hbar}[\,\hat{\mathcal{H}}_{\text{scatter}}, \hat{\sigma}]\,. \tag{3.106}$$

Here, $\nabla \hat{J}$ is the divergence of the spin current.

After some algebra, for details see [26], we get the following additional torques on the magnetization:

$$\mathbf{T} = \mathbf{S} \times (\mathbf{S} \times [(\mathbf{u} \cdot \nabla)\mathbf{S}]) + \beta \mathbf{S} \times [(\mathbf{u} \cdot \nabla)\mathbf{S}]\,. \tag{3.107}$$

3 The Equations of motion

The first term can be written as:

$$\mathbf{S} \times (\mathbf{S} \times [(\mathbf{u} \cdot \nabla)\mathbf{S}]) = \underbrace{(\mathbf{S} \cdot [(\mathbf{u} \cdot \nabla)\mathbf{S}])}_{=0}\mathbf{S} - \mathbf{S}^2\left[(\mathbf{u} \cdot \nabla)\mathbf{S}\right] = -\left[(\mathbf{u} \cdot \nabla)\mathbf{S}\right] . \tag{3.108}$$

The term $(\mathbf{S} \cdot [(\mathbf{u} \cdot \nabla)\mathbf{S}])$ on the right hand side vanishes because $(\mathbf{u} \cdot \nabla)\mathbf{S} \perp \mathbf{S}$. Taking the adiabatic as welll as the non-adiabatic spin torque term into account the following Gilbert equation:

$$\frac{\partial \mathbf{S}}{\partial t} = -\frac{\gamma}{\mu_s}\mathbf{S} \times \mathbf{H}_{\text{eff}} + \alpha \mathbf{S} \times \frac{\partial \mathbf{S}}{\partial t} - [(\mathbf{u} \cdot \nabla)\mathbf{S}] + \beta \mathbf{S} \times [(\mathbf{u} \cdot \nabla)\mathbf{S}] . \tag{3.109}$$

appears. The first two terms of this equation have been discussed before in section 3.2. These terms describe the precessional motion in the effective field $\mathbf{H}_{\text{eff}} = -\partial \mathcal{H}/\partial \mathbf{S}$ (first term) and the energy dissipation due to the damping (second term). The last two terms are the adiabatic (third term) and the non-adiabatic (fourth term) spin torque with the non-adiabaticity parameter β. $u = |\mathbf{u}|$ is given by

$$u = j_e P g \mu_B / \left(2 e M_s (1 + \beta^2)\right) ; \; M_s = \mu_s / a^3 , \tag{3.110}$$

with j_e being the current density, P the polarization, M_s the saturation magnetization.

As before [Eq. (3.87)], this equation can be brought to an explicit form. Therefore, the product $\mathbf{S}\times$ is usually added to both sides of Eq. (3.109) which leads to

$$\mathbf{S} \times \frac{\partial \mathbf{S}}{\partial t} = -\frac{\gamma}{\mu_s}\mathbf{S} \times (\mathbf{S} \times \mathbf{H}_{\text{eff}}) + \alpha \mathbf{S} \times \left(\mathbf{S} \times \frac{\partial \mathbf{S}}{\partial t}\right) \\ - \mathbf{S} \times [(\mathbf{u} \cdot \nabla)\mathbf{S}] + \beta \mathbf{S} \times (\mathbf{S} \times [(\mathbf{u} \cdot \nabla)\mathbf{S}]) . \tag{3.111}$$

Inserting the last expression into Eq. (3.109) and using the rule $\mathbf{a} \times (\mathbf{b} \times \mathbf{c}) = \mathbf{b}(\mathbf{a} \cdot \mathbf{c}) - \mathbf{c}(\mathbf{a} \cdot \mathbf{b})$:

$$\mathbf{S} \times \left(\mathbf{S} \times \frac{\partial \mathbf{S}}{\partial t}\right) = \underbrace{\left(\mathbf{S} \cdot \frac{\partial \mathbf{S}}{\partial t}\right)}_{=0} - \mathbf{S}^2 \frac{\partial \mathbf{S}}{\partial t} = -\frac{\partial \mathbf{S}}{\partial t} , \tag{3.112}$$

(the first term on the right hand side vanishes because of the orthogonality $\frac{\partial \mathbf{S}}{\partial t} \perp \mathbf{S}$), one gets

$$(1 + \alpha^2)\frac{\partial \mathbf{S}}{\partial t} = -\frac{\gamma}{\mu_s}\mathbf{S} \times [\mathbf{H}_{\text{eff}} + \alpha(\mathbf{S} \times \mathbf{H}_{\text{eff}})] \\ - (\mathbf{u} \cdot \nabla)\mathbf{S} - (\alpha - \beta)\mathbf{S} \times [(\mathbf{u} \cdot \nabla)\mathbf{S}] \\ + \alpha\beta\mathbf{S} \times (\mathbf{S} \times [(\mathbf{u} \cdot \nabla)\mathbf{S}]) . \tag{3.113}$$

3.3 The Landau-Lifshitz-Gilbert-Slonczewski equation

This equation can be simplified to:

$$\mathbf{S} \times (\mathbf{S} \times [(\mathbf{u} \cdot \nabla)\mathbf{S}]) = \underbrace{(\mathbf{S} \cdot [(\mathbf{u} \cdot \nabla)\mathbf{S}])}_{=0}\mathbf{S} - \mathbf{S}^2[(\mathbf{u} \cdot \nabla)\mathbf{S}]$$
$$= -[(\mathbf{u} \cdot \nabla)\mathbf{S}] . \qquad (3.114)$$

Here, we have used the fact that $(\mathbf{u} \cdot \nabla)\mathbf{S} \perp \mathbf{S}$ holds. Finally, one gets the modified Landau-Lifshitz-Gilbert (LLG) equation [25]:

$$\begin{aligned}\frac{\partial \mathbf{S}}{\partial t} = &- \frac{\gamma}{(1+\alpha^2)\mu_s}\mathbf{S} \times \mathbf{H}_{\text{eff}} - \frac{\alpha\gamma}{(1+\alpha^2)\mu_s}\mathbf{S} \times (\mathbf{S} \times \mathbf{H}_{\text{eff}}) \\ &- \frac{\alpha-\beta}{(1+\alpha^2)}\mathbf{S} \times [(\mathbf{u} \cdot \nabla)\mathbf{S}] \\ &+ \frac{1+\alpha\beta}{(1+\alpha^2)}\mathbf{S} \times (\mathbf{S} \times [(\mathbf{u} \cdot \nabla)\mathbf{S}]) , \end{aligned} \qquad (3.115)$$

which is identical to the "usual" LLG equation in the limit of $u \to 0$. This equation is sometimes called Landau-Lifshitz-Gilbert-Slonczewski equation.

The first two terms describe the precessional motion as well as the relaxation process due to the effective field $\mathbf{H}_{\text{eff}}$. These terms are identical to the "usual" Landau-Lifshitz-Gilbert equation (3.85). The last two terms describe the effect of the electric current. The third term describes a precessional motion while the fourth term describes a relaxation process. Here, we have to mention that it makes no sense anymore to speak about adiabatic and non-adiabatic spin torque because the nonadiabaticity parameter β occurs now in both terms.

Micromagnetic description of the spin torque terms

In micromagnetism it makes more sense to use the Gilbert equation (3.87) instead of the Landau-Lifshitz-Gilbert equation (3.115). The micromagnetic description of the first two "usual" terms of the Gilbert equation have been given before [see Eq. (3.120) and (3.121)]. Here, the description shall be restricted to a current in x-direction: $[\mathbf{u} \cdot \nabla]\mathbf{S} = u_x d\mathbf{S}/dx$. Furthermore, ϕ shall be the same for all positions in space (independent on x): $\phi = \phi(t)$. However, θ still depends on x: $\theta = \theta(x,t)$. Then, the spin itself is a function of θ and ϕ: $\mathbf{S} = \mathbf{S}(\theta, \phi)$.

For the spin torque terms we need:

$$\frac{d\mathbf{S}}{dx} = \frac{\partial \mathbf{S}}{\partial \theta}\frac{d\theta}{dx} = \frac{\partial \mathbf{S}}{\partial \theta}\frac{\partial \theta}{\partial x} . \qquad (3.116)$$

We know that in spherical coordinates $\mathbf{S} = \mathbf{e}_S$, and therefore:

$$\frac{d\mathbf{S}}{dx} = \frac{\partial \theta}{\partial x}\mathbf{e}_\theta + \sin\theta \underbrace{\frac{\partial \phi}{\partial x}}_{=0}\mathbf{e}_\phi = \frac{\partial \theta}{\partial x}\mathbf{e}_\theta . \qquad (3.117)$$

3 The Equations of motion

The last term on the right hand side vanishes because ϕ has been assumed to be independent of x.

Alternatively, we can use the right hand side of (3.116):

$$\frac{\partial \mathbf{S}}{\partial \theta}\frac{\partial \theta}{\partial x} = \left(\underbrace{\frac{\partial \theta}{\partial \theta}}_{=1} \mathbf{e}_\theta + \sin\theta \underbrace{\frac{\partial \phi}{\partial \theta}}_{=0} \mathbf{e}_\phi \right) \left(\frac{\partial \theta}{\partial x} \right) = \frac{\partial \theta}{\partial x}\mathbf{e}_\theta . \qquad (3.118)$$

ϕ is also independent of θ.

Furthermore, we need:

$$\mathbf{S} \times \frac{\mathrm{d}\mathbf{S}}{\mathrm{d}x} = \left(\frac{\partial \theta}{\partial x} \right)(\mathbf{e}_S \times \mathbf{e}_\theta) = \frac{\partial \theta}{\partial x}\mathbf{e}_\phi . \qquad (3.119)$$

Then, we can write down the Gilbert equation (3.87) as:

$$\dot\theta = -\alpha \sin\theta \dot\phi + \frac{\gamma}{M_S \sin\theta}\frac{\delta \mathcal{E}}{\delta \phi} - u_x \frac{\partial \theta}{\partial x} \qquad (3.120)$$

$$\dot\phi = \frac{\alpha \dot\theta}{\sin\theta} - \frac{\gamma}{M_S \sin\theta}\frac{\delta \mathcal{E}}{\delta \theta} + \frac{\beta u_x}{\sin\theta}\frac{\partial \theta}{\partial x} . \qquad (3.121)$$

For the energy density $\mathcal{E}$:

$$\mathcal{E} = A\left(\frac{\partial \theta}{\partial x}\right)^2 - K\cos^2\theta + K_h \sin^2\theta \sin^2\phi - M_S B\cos\theta , \qquad (3.122)$$

we get explicitly:

$$\dot\theta = -\alpha \sin\theta \dot\phi + \frac{K_h \gamma}{M_S}\sin\theta \sin(2\phi) - u_x \frac{\partial \theta}{\partial x} \qquad (3.123)$$

$$\dot\phi = \frac{\alpha \dot\theta}{\sin\theta} - \frac{2\gamma}{M_S}\left[\cos\theta\left(K + K_h \sin^2\phi\right) - \frac{A}{\sin\theta}\frac{\partial^2 \theta}{\partial x^2}\right] - \gamma B + \frac{\beta u_x}{\sin\theta}\frac{\partial \theta}{\partial x} . \qquad (3.124)$$

The second term $A\sin^2\theta\left(\frac{\partial \phi}{\partial x}\right)^2$ of the exchange interaction $\mathcal{E}_J$ [see Eq. (2.20)] has been skipped due to the assumption that ϕ does not depend on x, therefore $\frac{\partial \phi}{\partial x} = 0$.

Eq. (3.123) and (3.124) are the differential equations we have to solve if we want to consider the dynamics influenced by an electric current u_x, respectively an external field B.

3.4 Numerical method to solve the equation of motion

Independent of whether we want to solve the Landau-Lifshitz, the Gilbert, or the Landau-Lifshitz-Gilbert equation we can do it analytically or with aid of a computer (numerical simulations). However, due to the complexity of the differential equations we are restricted to some simple situations in the case of analytical calculations. All these equations are non-linear vector differential equations of the first order. This means we have to solve two (spherical coordinates) or three (cartesian coordinates) coupled nonlinear differential equations. Especially the nonlinearity makes it difficult to solve these differential equations.

Numerics gives us the chance to solve these differential equations also in complex cases which cannot be solved analytically. The goal of this section is to describe a numerical procedure to deal with a given equation of motion for the classical spin dynamics.

Some first remarks

First, we have to decide which coordinate system we want to use for our description. Depending on whether we choose the sperical or the cartesian coordinate system we have three or two differential equations. So it seems to be clear that the sperical coordinate system gives us an advantage. Here, we have to solve only N times two coupled differential equations for θ_n and ϕ_n in the case of N spins. This assumption is deceptive. Using cartesian coordinates will lead to a faster performance of the program. Here, we have to solve three times N differential equations for S_n^x, S_n^y, and S_n^z, however we have not to deal with trigonometric functions. The numerical effort of mathematical operations are not the same, e.g. multiplications are faster than divisions. Therefore, it makes sense to use $0.5*x$ instead of $x/2$ in the program. The same is true for trigonometric functions. Trigonometric functions reduce the performance speed of the program. In this case the compiler has to deal with series expansions instead of a normal number. Therefore, our decision should be to solve the equation of motion in cartesian coordinates.

Then, we have to think about the dimension of our equation of motion. Here, it shall be discussed at the example of the Landau-Lifshitz-Gilbert equation:

$$\frac{(1+\alpha^2)\,\mu_S}{\gamma}\frac{d\mathbf{S}_n}{dt} = -\mathbf{S}_n \times \left(\mathbf{H}_{eff}^n + \alpha \mathbf{S}_n \times \mathbf{H}_{eff}^n\right)\,. \qquad (3.125)$$

As discussed before, the spins $\mathbf{S}_n$ as well as the Gilbert damping α are dimensionless. The effective field $\mathbf{H}_{eff}^n$ has the same dimension as the Hamilton function $\mathcal{H}$: Joule. The other constants have the dimension: $[\gamma] = 1/(Ts)$ and $[\mu_S] = J/T$. This means that both sides of the Landau-Lifshitz-Gilbert equation have the dimension Joule.

3 The Equations of motion

For numerical simulations it is quite common to use dimensionless coordinates. To do so we have to devide both sides of the Landau-Lifshitz-Gilbert equation with a constant which has the dimension Joule, e.g. the exchange constant J. Then we can write the dimensionless Landau-Lifshitz-Gilbert equation as:

$$\left(1+\alpha^2\right)\frac{\mathrm{d}\mathbf{S}_n}{\mathrm{d}\tau} = -\mathbf{S}_n \times \left(\mathbf{h}_{eff}^n + \alpha \mathbf{S}_n \times \mathbf{h}_{eff}^n\right), \qquad (3.126)$$

with the dimensionless time $\tau = J\gamma t/\mu_S$ and the dimensionless effective field $\mathbf{h}_{eff}^n = -1/J \partial \mathcal{H}/\partial \mathbf{S}_n$.

The third point we have to think about is the time step within our simulation. If we solve our equation of motion with the aid of a numerical method we have to use discrete time steps $t_n = n\Delta t$, $n \in \mathbb{N}$.

The time steps have to be small to reduce the numerical error and to reproduce the correct physics. On the other hand, a smaller time step means that we need more time steps to simulate the same distance in time, and this means that the numerical effort increases. To know the correct Δt we have to think about the intrinsic times in our system. In our case the magnetic moment can precess (Larmor precession) and relax. The relaxation time has been calculated in Sec. 3.2: Eq. (3.81). The precessional time can be estimated by $\tau_\mathrm{p} = \alpha \tau_\mathrm{rel}$. Alternatively, we can calculate the precession time from the Larmor frequency. In the case of no damping $\alpha = 0$, the Larmor frequency is given by

$$\omega_L = \frac{\gamma}{\mu_S}|\mathbf{H}_{eff}|. \qquad (3.127)$$

With damping we can approximate ω_L to

$$\omega_L = \frac{\gamma}{\mu_S\left(1+\alpha^2\right)}|\mathbf{H}_{eff}|. \qquad (3.128)$$

This means the Larmor frequency decreases with increasing damping. This makes sense because it means that the precession becomes slower and the precession time $T = 2\pi/\omega_L$ increases. The slowest precession time T appears for $\alpha = 0$:

$$T = \frac{2\pi\mu_S}{\gamma|\mathbf{H}_{eff}|}. \qquad (3.129)$$

To get the correct dynamics, one time step Δt has to be 100-1000 times smaller than the precession time T to reproduce the precessional motion of the spin.

The Heun method

So far, we have discussed just some basics. Let us come to the method which has been used during the studies: the Heun method. The Heun method is a predictor-corrector method. This means this method does not immediately calculate the new magnetization as in the case of the Euler method. Due to this additional predictor step the Heun method achieves a higher accuracy as the Euler method. The Heun method first calculates an intermediate value of the magnetization:

$$\overline{S}_\eta(t_{n+1}) = S_\eta(t_n) + f_\eta\left[\mathbf{S}(t_n), \mathbf{H}_{\text{eff}}(t_n), t_n\right] \Delta t , \qquad (3.130)$$

before the final approximation at the next integration point will be calculated by:

$$S_\eta(t_{n+1}) = S_\eta(t_n) + \frac{1}{2}\left\{ f_\eta\left[\mathbf{S}(t_n), \mathbf{H}_{\text{eff}}(t_n), t_n\right] + f_\eta\left[\overline{\mathbf{S}}(t_{n+1}), \overline{\mathbf{H}}_{\text{eff}}(t_{n+1}), t_{n+1}\right] \right\} \Delta t . \qquad (3.131)$$

Here, $f_\eta\left[\mathbf{S}(t_n), \mathbf{H}_{\text{eff}}(t_n), t_n\right]$ is the η component ($\eta \in \{x, y, z\}$) of the right hand side of our equation of motion at time t_n.

3 The Equations of motion

4 Field and current driven domain wall motion

After introducing the Heisenberg model and explaining the underlying equations of motions, this thesis provides an overview of the dynamics of transverse domain walls (see Fig. 4.1). The presentation is restricted to transverse domain walls because in the case of transverse domain walls it is possible to compare the analytical descriptions with numerical simulations. This is not the case for complex domain walls like the vortex domain wall [27, 28]. In these cases we are mostly restricted to simulations. However, it is often the case that we can use the results and conclusions of the transverse domain walls to describe complex domain walls.

The following chapter introduces transverse domain walls and discusses field and current driven domain wall motion. For that, we use the micromagnetic formalism. As explained in chapter 2 micromagnetism is a classical description which can be used to describe magnetism on the microscopic level. For the numerical investigations we have used the classical Heisenberg model.

4.1 Static ferromagnetic domain walls

Starting point of all analytical (micromagnetic) calculations is the energy density $\mathcal{E}$. In Sec. 2 we have derived the energy resp. energy density corresponding to the Heisenberg Hamiltonian:

$$\mathcal{E} = A\left[\left(\frac{\partial \theta}{\partial x}\right)^2 + \sin^2\theta \left(\frac{\partial \phi}{\partial x}\right)^2\right] - K\cos^2\theta + K_h \sin^2\theta \sin^2\phi - B\cos\theta . \quad (4.1)$$

The first term describes the exchange interaction along the x-axis. In most cases the transverse domain wall is quasi-1D or 1D. Therefore, we can restrict our investigation to a linear chain. The extension to quasi-1D samples (e.g. transverse domain walls in cylindrical nanowires) is straightforward. Therefore, the domain wall solution will also become one-dimensional.

The second and third term are uniaxial anisotropies. In this case an easy axis anisotropy in z-direction and a hard axis anisotropy in y-direction has been assumed. The last term describes an external field in $+z$ direction.

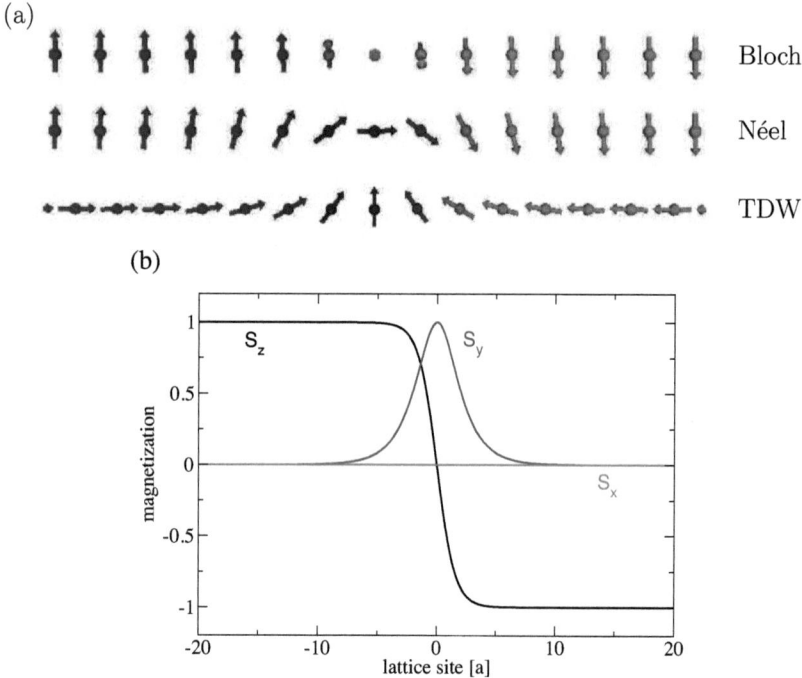

Figure 4.1: Different forms of domain walls (a) and transverse domain wall profile (b).

4.1 Static ferromagnetic domain walls

Transverse domain walls appear due to the competition of the exchange interaction A and the easy axis anisotropy K. The hard-axis anisotropy K_h determines the orientation of the domain wall in space. K_h is responsible for a fixed ϕ angle. However, to stabilize the domain wall we also need fixed ends with opposite orientation of the magnetic moments. On the macroscopic scale domain walls appear due to the competition of the exchange interaction and the long-ranged dipole-dipole interaction (not included here). In this case the fixed ends are not necessary to stabilize the domain wall.

The next step in the analytical description of domain walls is to solve the Gilbert equation:

$$\dot{\theta} = -\alpha \sin\theta \dot{\phi} + \frac{\gamma}{M_S \sin\theta} \frac{\delta\mathcal{E}}{\delta\phi} \tag{4.2}$$

and

$$\sin\theta \dot{\phi} = \alpha \dot{\theta} - \frac{\gamma}{M_S} \frac{\delta\mathcal{E}}{\delta\theta} . \tag{4.3}$$

At first, we are interested in static domain wall solutions. This means $\dot{\theta} = \dot{\phi} = 0$, i.e. no time dependence of the angles. Furthermore, $B = 0$. The external field would lead to a domain wall motion as we will see later.

With this in mind we immediately see that the Gilbert equation reduces to the search of the lowest energy configuration. Or, with other words: we have to calculate just the variations:

$$\frac{\delta\mathcal{E}}{\delta\phi} = 2K_h \sin^2\theta \sin\phi \cos\phi - 2A \frac{\partial}{\partial x}\left(\sin^2\theta \frac{\partial\phi}{\partial x}\right) = 0 \tag{4.4}$$

$$\frac{\delta\mathcal{E}}{\delta\theta} = 2\sin\theta \cos\theta \left(K + K_h \sin^2\phi + A\left(\frac{\partial\phi}{\partial x}\right)^2\right) - 2A \frac{\partial^2\theta}{\partial x^2} = 0 . \tag{4.5}$$

To solve these equations we have to make an additional assumption: Here, we are searching for a solution with $\phi(x,t) = \varphi = \text{const.}$. This means, we assume that the domain wall is just characterized by the angle θ. Otherwise the calculation becomes too complex to calculate analytically. This assumption also means that $\dot{\phi} = \phi' = 0$.

Therefore, Eq. (4.4) has no important meaning anymore and Eq. (4.5) becomes:

$$A \frac{\partial^2 \theta}{\partial x^2} = \left(K + K_h \sin^2\varphi\right) \sin\theta \cos\theta . \tag{4.6}$$

The solution of this second order differential equation will give us the shape of the domain wall. The first step in solving this differential equation is to multiply both sides with $\frac{\partial \theta}{\partial x}$ and integrating over x. This leads to the first order differential equation:

$$A \left(\frac{\partial \theta}{\partial x} \right)^2 = - \left(K + K_h \sin^2 \varphi \right) \cos^2 \theta + C \, . \qquad (4.7)$$

C is an integration constant which has to be determined by the boundary conditions at $x = \pm\infty$. At this point we can assign the boundary conditions and try to solve the corresponding differential equation. However, it shall be demonstrated that the differential equation (4.7) can be solved also without knowing the correct boundary condition. In this case we get a general solution which can be substantiated at the end by assigning the boundary conditions.

On the next step, we have to define some abbreviations:

$$\Delta = \sqrt{\frac{A}{K + K_h \sin^2 \varphi}} \, , \qquad (4.8)$$

and

$$\kappa = \sqrt{\frac{K + K_h \sin^2 \varphi}{C}} \, . \qquad (4.9)$$

We will see later that Δ corresponds to the domain wall width and κ covers the boundary condition. With these abbreviations we can write Eq. (4.7) as:

$$\Delta \kappa \left(\frac{\partial \theta}{\partial x} \right) = \sqrt{1 - \kappa^2 \cos^2 \theta} \, . \qquad (4.10)$$

This differential equation can easily be solved via the separation of variables:[1]

$$\int_0^\theta \frac{d\theta'}{\sqrt{1 - \kappa^2 \cos^2 \theta'}} = \frac{1}{\Delta \kappa} \int_0^x dx' \, . \qquad (4.11)$$

To integrate the left hand side it is helpful to make the following substitution: $\theta \to \theta - \frac{\pi}{2} \Rightarrow \cos \theta \to \cos \left(\theta - \frac{\pi}{2} \right) = \sin \theta$. This substitution is just a change from the definition of spherical coordinates with $S_z = S \cos \theta$ to the second definition with $S_z = S \sin \theta$. Due to this substitution the right hand side of Eq. (4.11) is now identical to the definition of the elliptic integral of the first kind $F(\kappa, \theta)$:

$$F(\kappa, \theta) = \int_0^\theta \frac{d\theta'}{\sqrt{1 - \kappa^2 \sin^2 \theta'}} = \frac{x}{\Delta \kappa} \, . \qquad (4.12)$$

---

[1] The $'$ behind θ and x just means that the integration variables and the limits of the integrals are different: θ' and x' are variables and θ and x the integration limits.

4.1 Static ferromagnetic domain walls

With Eq. (4.12), we have found a general solution of the differential equation (4.6).

From mathematics it is known that:

$$\theta = F^{-1}(\kappa, \theta) = \operatorname{am}(u, \kappa) , \tag{4.13}$$

where $F^{-1}(\kappa, \theta)$ is the inverse elliptical integral of first kind and $\operatorname{am}(u, \kappa)$ is the Jacobi amplitude. Within $\operatorname{am}(u, \kappa)$ u is given by:

$$u = F(\kappa, \theta) = \frac{x}{\Delta\kappa} . \tag{4.14}$$

Then we are able to write down the z-component, respectively the perpendicular ($\perp$)-component of the domain wall as:

$$S_z = \sin\theta = \operatorname{sn}(u, \kappa) = \operatorname{sn}\left(\frac{x}{\Delta\kappa}, \kappa\right) \tag{4.15}$$

$$S_\perp = \cos\theta = \operatorname{cn}(u, \kappa) = \operatorname{cn}\left(\frac{x}{\Delta\kappa}, \kappa\right) . \tag{4.16}$$

$\operatorname{sn}(u, \kappa)$ and $\operatorname{cn}(u, \kappa)$ are the Jacobi sine resp. Jacobi cosine.

It should be noticed that this are solutions in the transformed space, because we have exchanged $\cos\theta$ by $\sin\theta$. To get the correct solution we have to make a backtransformation to the original coordinate system. During this backtransformation $\sin\theta$ becomes $\cos\theta$ and $\cos\theta$ becomes $-\sin\theta$. However, we can ignore the sign change for $S_\perp$ and therefore the final solution becomes:

$$S_z = \cos\theta = \operatorname{sn}\left(\frac{x}{\Delta\kappa}, \kappa\right) \tag{4.17}$$

$$S_\perp = \sin\theta = \operatorname{cn}\left(\frac{x}{\Delta\kappa}, \kappa\right) . \tag{4.18}$$

Now, we have to think about the boundary conditions. For $x \to \pm\infty$ which corresponds to $\theta \to \pm\pi = \text{const.}$ and $\partial\theta/\partial x \to 0$ the integration constant becomes $C = K + K_h \sin^2\varphi$ [see Eq. (4.7)] and therefore $\kappa = 1$. In this limit the domain wall solution is given by:

$$S_z = \operatorname{sn}\left(\frac{x}{\Delta}, 1\right) = \tanh\left(\frac{x}{\Delta}\right) \tag{4.19}$$

$$S_\perp = \operatorname{cn}\left(\frac{x}{\Delta}, 1\right) = \operatorname{sech}\left(\frac{x}{\Delta}\right) . \tag{4.20}$$

Depending on the environment and the exact definition of ϕ this domain wall solution is called Bloch, Néel, or transverse domain wall (see Fig. 4.1).

In micromagnetism we could also find another description of the domain wall given by the angle θ. We can get this formula from $S_z = \cos\theta$ together with Eq. (4.19):

$$\theta = \arccos\left(\pm\tanh\left(\frac{x}{\Delta}\right)\right) = \arctan\left(e^{\pm\frac{x}{\Delta}}\right) . \tag{4.21}$$

4 Field and current driven domain wall motion

Another interesting limit appears if the domain wall is constricted. This situation corresponds to $\partial\theta/\partial x \to \infty$ for $x \to \pm\infty$. In this case it is easy to show that $C \to \infty$ and $\kappa \approx 0$. Furthermore, we see that in this cases the domain wall width Δ is given by the length of the constriction: $\Delta = L$. Therefore, the domain wall is described by:

$$S_z = \operatorname{sn}\left(\frac{x}{L}, 0\right) = \cos\left(\frac{x}{L}\right) \tag{4.22}$$

$$S_\perp = \operatorname{cn}\left(\frac{x}{L}, 0\right) = \sin\left(\frac{x}{L}\right). \tag{4.23}$$

A few words about the domain wall energy: Starting point of the calculation is the system energy:

$$E = \int_{-\infty}^{+\infty} \left(A\left[\left(\frac{\partial\theta}{\partial x}\right)^2 + \cos^2\theta \left(\frac{\partial\phi}{\partial x}\right)^2\right] + K\cos^2\theta + K_h \cos^2\theta \sin^2\phi \right) dx. \tag{4.24}$$

This energy is similar to Eq. (4.1). However, we are using a spherical coordinate system which is defined by $-\pi/2 \le \theta \le +\pi/2$. This is similar to the transformation $\theta \to \theta + \pi/2$ which we have made before. The reason is again the definition of the elliptic integrals. Furthermore, the Zeeman term has been skipped and the easy axis anisotropy is defined as an effective hard axis anisotropy to simplify the calculation.

Therefore, we can define the effective anisotropy:

$$K_{\text{eff}} = K + K_h \sin^2\phi. \tag{4.25}$$

After a similar calculation as before we get from the variation $\delta\mathcal{E}/\delta\theta$:

$$\frac{\partial\theta}{\partial x} = \frac{\sqrt{1 - \kappa^2 \sin^2\theta}}{\kappa\Delta}, \tag{4.26}$$

with Δ and κ given by Eq. (4.8) resp. Eq. (4.9). With this equation it is quite easy to write down the domain wall energy as:

$$\begin{aligned} E &= \sqrt{AK_{\text{eff}}} \int_{-\frac{\pi}{2}}^{+\frac{\pi}{2}} \left(\frac{\sqrt{1 - \kappa^2 \sin^2\theta}}{\kappa} + \frac{\kappa \cos^2\theta}{\sqrt{1 - \kappa^2 \sin^2\theta}} \right) d\theta \\ &= \sqrt{AK_{\text{eff}}} \left(\frac{2}{\kappa} E\left(\kappa, \frac{\pi}{2}\right) + \frac{\kappa^2 - 1}{\kappa} F\left(\kappa, \frac{\pi}{2}\right) \right). \end{aligned} \tag{4.27}$$

$F\left(\kappa, \frac{\pi}{2}\right)$ and $E\left(\kappa, \frac{\pi}{2}\right)$ are the elliptical integrals of first and second order. In the limit of a normal 180° domain wall with $\kappa = 1$ the energy becomes:

$$E = 2\sqrt{AK_{\text{eff}}} \int_{-\frac{\pi}{2}}^{+\frac{\pi}{2}} \cos\theta \, d\theta = 4\sqrt{AK_{\text{eff}}} \,. \tag{4.28}$$

4.2 Analytical considerations of the dynamics: the q-ϕ model

So far, we have calculated the profile of a transverse domain wall. Therefore, we had to set $B = 0$. The goal is now to describe the dynamics and to write down the velocity equations under the influence of an external magnetic field ($B \neq 0$) and a spin-polarized electric current. A model which can exactly describe the domain wall dynamics in a simple way is the q-ϕ model. This model shall be described in the following.

The description starts with the domain wall solution Eq. (4.21):

$$\theta(x,t) = 2\arctan\left(e^{-\frac{x-q(t)}{\Delta}}\right) \,. \tag{4.29}$$

The only difference here is that we assume that the domain wall is already moving with a constant velocity. Therefore, the position of the domain wall (center) becomes time dependent which is expressed by the additional factor $q(t)$.

From the domain wall profile Eq. (4.29) we see immediately that the following conditions hold:

$$\frac{\partial \theta}{\partial x} = -\frac{\sin\theta}{\Delta}, \tag{4.30}$$

$$\frac{\partial^2 \theta}{\partial x^2} = \frac{\partial}{\partial x}\left(-\frac{\sin\theta}{\Delta}\right) = -\frac{\cos\theta}{\Delta}\frac{\partial \theta}{\partial x} = -\frac{\cos\theta \sin\theta}{\Delta^2}, \tag{4.31}$$

and for the velocity of the domain wall:

$$v = \dot{q} = \frac{\Delta}{\sin\theta}\dot{\theta} \,. \tag{4.32}$$

The second condition can be easily proven by differentiating θ two times with respect to x and taking into account that $\sin\theta = \text{sech}\left(-\frac{x-q(t)}{\Delta}\right)$ and $\cos\theta = \tanh\left(-\frac{x-q(t)}{\Delta}\right)$.

4 Field and current driven domain wall motion

With the first condition we can immediately write the Gilbert equations (3.120) and (3.121) together with the energy density Eq. (4.1) as:

$$\dot{\theta} = -\alpha \sin\theta \dot{\phi} + \frac{K_h \gamma}{M_S} \sin\theta \sin(2\phi) + u_x \frac{\sin\theta}{\Delta} \qquad (4.33)$$

$$\dot{\phi} = \frac{\alpha \dot{\theta}}{\sin\theta} - \frac{2\gamma \cos\theta}{M_S}\left[\left(K + K_h \sin^2\phi\right) - \frac{A}{\Delta^2}\right] - \gamma B - \frac{\beta u_x}{\Delta}. \qquad (4.34)$$

The trick of the q-ϕ model is now to assume that we are in the center of the domain wall. In the domain wall center we have $\theta = \pi/2$. Therefore, the differential equations (4.33) and (4.34) become simplified:

$$\dot{\theta} = -\alpha\dot{\phi} + \frac{K_h \gamma}{M_S}\sin(2\phi) + \frac{u_x}{\Delta} \qquad (4.35)$$

$$\dot{\phi} = \alpha\dot{\theta} - \gamma B - \frac{\beta u_x}{\Delta}. \qquad (4.36)$$

The goal now is to solve these differential equations under different boundary conditions. In the following we will discuss the different conditions which correspond to the two known types of motion: domain wall motion with a direct reversal $\dot{\phi} = 0$ and the domain wall motion where we see an oscillation during the reversal process $\dot{\phi} \neq 0$. Within the first situation $\dot{\phi} = 0$ (direct reversal) we have to devide further into two scenarios which lead to different velocities. The direct reversal can appear due to an additional hard axis anisotropy K_h or due to some other effects. In the latter case we assume $\dot{\phi} = 0$ and at the same time we set $K_h = 0$. Such a situation occurs in the cases of vortex domain walls in cylindrical nanowires [27, 28]. In these cases the inner shape of the vortex domain wall, the vortex itself, leads to a direct reversal. An oscillation during the domain wall motion would lead to a destruction of the vortex which would lead to an increase of the energy. Therefore, this reversal process will not appear.

In the following, we will restrict ourself to transverse domain walls. However, we will take into account all possible and impossible scenarios. This means we will also discuss the scenario with $\dot{\phi} = 0$ and $K_h = 0$ even if it is unlikely in the case of a transverse domain wall, as we will see in the following.

I Domain wall motion with direct reversal $\dot{\phi} = 0$ and $K_h = 0$

Under these conditions the differential equations (4.35) and (4.36) become:

$$\dot{\theta} = \frac{u_x}{\Delta} \qquad (4.37)$$

$$0 = \alpha\dot{\theta} - \gamma B - \frac{\beta u_x}{\Delta}. \qquad (4.38)$$

4.2 Analytical considerations of the dynamics: the q-ϕ model

With the assumption of being in the center of the domain wall ($\theta = \pi/2$) Eq. (4.32) becomes $\dot{\theta} = v/\Delta$. Inserting this into (4.37) leads immediately to the domain wall velocity:

$$v = u_x . \qquad (4.39)$$

With the same trick we get from Eq. (4.38):

$$v = \frac{\gamma B}{\alpha}\Delta + \frac{\beta}{\alpha}u_x . \qquad (4.40)$$

Here, we have solved two independent differential equations. Therefore, we get two independent solutions which fulfill the condition $\dot{\phi} = 0$ at the same time. Therefore, we can simply say that the conditions $\dot{\phi} = 0$ and $K_h = 0$ do not lead to a unique solution. However, we will see that both solutions have a physical relevance.

The first equation Eq. (4.37) does not contain the external field B, therefore the solution Eq. (4.39) does not contain B either. Eq. (4.38) contains the external field and therefore the corresponfing solution Eq. (4.40) does so, too. However, we are not able yet to get a unique solution. Here, we need more information or a unique boundary condition. This is, for instance, the case if we choose as boundary condition: $\dot{\phi} = 0$ and $K_h \neq 0$.

II Domain wall motion with direct reversal $\dot{\phi} = 0$ and $K_h \neq 0$

Under these conditions the differential equations (4.35) and (4.36) become:

$$\dot{\theta} = \frac{K_h \gamma}{M_S}\sin(2\phi) + \frac{u_x}{\Delta} \qquad (4.41)$$

$$0 = \alpha\dot{\theta} - \gamma B - \frac{\beta u_x}{\Delta} . \qquad (4.42)$$

Then, we plug Eq. (4.41) into Eq. (4.42). This leads to a condition for ϕ:

$$\phi = \frac{1}{2}\arcsin\left(\frac{M_S B + (\beta - \alpha)\frac{u_x M_S}{\gamma\Delta}}{\alpha K_h}\right) . \qquad (4.43)$$

This formula determines the stability of the domain wall motion. A change of ϕ leads to a change of the domain wall width Δ [see Eq. (4.8)] as well as the domain wall energy E_{DW} [see Eq. (4.28)]. If the angle ϕ becomes $90°$ we have the highest domain wall energy. In this case the spins inside the domain wall are in the direction of the hard axis anisotropy K_h. A further increase of the strength of the external field B or the current u_x or a decrease of the Gilbert damping α

4 Field and current driven domain wall motion

leads to the fact that the spins can overcome the energy barrier given by K_h and the domain wall starts to precess during the motion. This phenomenon is called Walker breakdown and corresponds to a decrease of the velocity. In the case of field driven domain wall motion it is known that domain walls with a precessional motion are always slower than domain walls with a direct reversal [29]. In the case of current driven domain wall motion this is not necessarily the case [25]. However, in most of the cases it is also true.

With $\dot{\theta} = v/\Delta$ (third condition Eq. (4.32) with $\theta = \pi/2$) in Eq. (4.41) we get finally the velocity:

$$v = \frac{\gamma B}{\alpha}\Delta + \frac{\beta}{\alpha}u_x \ . \tag{4.44}$$

This scenario shows that the boundary conditions $\dot{\phi} = 0$ and $K_h \neq 0$ lead to a unique solution. The additional hard axis anisotropy $K_h \neq 0$ breaks the symmetry in the energy landscape and leads to a direct reversal.

The next scenario assumes a domain wall motion where we have a precessional motion $\dot{\phi} \neq 0$ due to $K_h = 0$.

III Domain wall motion with oscillation

The boundary conditions here are $K_h = 0$ and $\dot{\phi} \neq 0$. Under these conditions the differential equations (4.35) and (4.36) become:

$$\dot{\theta} = -\alpha\dot{\phi} + \frac{u_x}{\Delta} \tag{4.45}$$

$$\dot{\phi} = \alpha\dot{\theta} - \gamma B - \frac{\beta u_x}{\Delta} \ . \tag{4.46}$$

Inserting (4.46) in (4.45) and using the condition $v = \dot{\theta}\Delta$ we get:

$$v = \frac{\alpha\gamma B}{1 + \alpha^2}\Delta + \frac{1 + \alpha\beta}{1 + \alpha^2}u_x \ . \tag{4.47}$$

Then, with

$$\frac{\alpha}{1 + \alpha^2} = \frac{1}{\alpha + \frac{1}{\alpha}} \tag{4.48}$$

we can finally write:

$$v = \frac{\gamma B}{\alpha + \frac{1}{\alpha}}\Delta + \frac{1 + \alpha\beta}{1 + \alpha^2}u_x \ . \tag{4.49}$$

Please notice: in the case $\alpha = \beta$ the "current contribution" of the velocity v becomes:

$$v_u = u_x . \tag{4.50}$$

The same is true for v_u in the scenario (II):

$$v_u = \frac{\beta}{\alpha} u_x \xrightarrow{\alpha=\beta} u_x . \tag{4.51}$$

Inserting (4.45) in (4.46) leads to:

$$\dot{\phi} = \frac{\alpha - \beta}{1 + \alpha^2} \frac{u_x}{\Delta} - \frac{\gamma B}{1 + \alpha^2} . \tag{4.52}$$

This is the precession frequency of the domain wall $\omega = \dot{\phi} = v/r$, where r has the dimension of a length. With the characteristic length, the domain wall width Δ, the precession velocity can be written as:

$$v_{\text{prec.}} = \dot{\phi}\Delta = \frac{\alpha - \beta}{1 + \alpha^2} u_x - \frac{\gamma B}{1 + \alpha^2} \Delta . \tag{4.53}$$

Final remarks

The previous calculation is based on the profile given by Eq. (4.29). The question is now, what happens if we use instead the profile:

$$\theta(x,t) = 2\arctan\left(e^{\frac{x-q(t)}{\Delta}}\right) . \tag{4.54}$$

The only difference is the sign in the exponential function which has changed. The result of this sign change can be seen in Fig. 4.2. The domain where all the spins are in $+z$-direction ($\theta = 0$) is now on the left side of the domain wall.

In this case we find the following conditions:

$$\frac{\partial \theta}{\partial x} = \frac{\sin \theta}{\Delta} , \tag{4.55}$$

and

$$v = \dot{q} = -\frac{\Delta}{\sin \theta} \dot{\theta} , \tag{4.56}$$

and with the assumption of being in the center of the domain wall $\theta = \pi/2$:

$$\frac{\partial \theta}{\partial x} = \frac{1}{\Delta} , \tag{4.57}$$

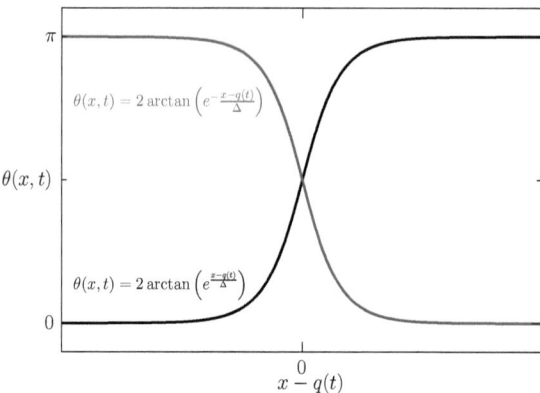

Figure 4.2: θ-profile of the domain wall

and

$$v = -\Delta \dot{\theta} \ . \tag{4.58}$$

If we make the same considerations as before, just with the conditions Eq. (4.57), and Eq. (4.58) we get the following velocities of the domain wall:

$$v_{\text{Ia}} = -u_x \ , \tag{4.59}$$
$$v_{\text{Ib}} = -\frac{\gamma B}{\alpha}\Delta - \frac{\beta}{\alpha}u_x \ , \tag{4.60}$$
$$v_{\text{II}} = v_{\text{Ib}} \ , \tag{4.61}$$
$$v_{\text{III}} = -\frac{\gamma B}{\alpha + \frac{1}{\alpha}}\Delta - \frac{1+\alpha\beta}{1+\alpha^2}u_x \ . \tag{4.62}$$

These are the same velocities as before just with a negative sign. This means the domain wall moves now in the opposite direction.

For the precessional velocity in scenario III we find with the new profile:

$$v_{\text{prec.}} = -\frac{\alpha - \beta}{1+\alpha^2}u_x - \frac{\gamma B}{1+\alpha^2}\Delta \ . \tag{4.63}$$

In this case the field term (second term) is unchanged but the current term (first term) has changed its sign.

How can we explain these results? In the case of the external field this is quite easy. We haven't changed the energy terms in Eq. (4.1). Therefore, the external

4.2 Analytical considerations of the dynamics: the q-ϕ model

field is still in $+z$-direction. However, we have changed the profile. The domain with the parallel spin alignment with respect to the external field has changed the side. The Zeeman term (external field) leads to a lower energy if the domain with the parallel orientation of spin and external field becomes larger. This means the two domain walls described by the two profiles Eq. (4.29) and Eq. (4.54), have to move in opposite directions.

Let us come now to the current contribution. In Sec. 3.3 we have seen that $\partial\theta/\partial x$ is the responsible condition for the current contribution in a micromagnetic description. Due to the change of the sign in the profile this condition also changes its sign. Nevertheless, these are mathematical details and do not answer the question: "How can we explain these results?". The explanation or conclusion (depending on how we want to see it) is the following. In the derivation of the micromagnetic description of the spin torque term (Sec. 3.3) we have assumed that the current always flows from the "up" domain (spins in $+z$-direction, $\theta = 0$) to the "down" domain (spins in $-z$-direction, $\theta = \pi$). This leads to a change of the moving direction if we change the profile, and therefore, we get the negative velocities. This problem with the destinguished current flow is also included in the simulations. To change the direction of the current flow, we have to change the sign of **u** in Eq. (3.115).

Domain wall mass

It is known that the domain wall can be seen as a quasi particle with a certain mass m [30, 29]. Within the q-θ model it is quite easy to estimate this mass.

Under the assumption of a static domain wall $u_x = B = 0$ and $\alpha = 0$, we can write the first of the Gilbert equations with the boundary condition $\dot\phi \neq 0$ and $K_h \neq 0$ Eq. (4.41) as:

$$\dot q = 2\frac{K_h \gamma}{M_S} \Delta \sin\phi \cos\phi \,. \tag{4.64}$$

Here, we have used the condition $\dot q = v = \dot\theta \Delta$. Then, the kinetic energy E_{kin} of the domain wall is given by:

$$E_{\text{kin}} = \frac{1}{2}m\dot q = 2m\frac{K_h^2 \gamma^2}{M_S^2}\Delta^2 \sin^2\phi \cos^2\phi \,. \tag{4.65}$$

We know the energy of the system [see Eq. (4.24)]. However, we only need the contribution of the hard axis anisotropy K_h:

$$E_{K_h} = \int_{-\infty}^{+\infty} K_h \sin^2\theta \sin^2\phi \, \mathrm{d}\theta \,. \tag{4.66}$$

4 Field and current driven domain wall motion

With $\sin^2\phi = (1 - \sqrt{1 - 4\sin^2\phi\cos^2\phi})/2$, and Eq. (4.65) we get:

$$E_{K_h} = C \int_{-\infty}^{+\infty} \sin^2\theta \, d\theta = C \int_{-\infty}^{+\infty} \text{sech}\left(\frac{x}{\Delta}\right) dx = C\Delta \tanh\left(\frac{x}{\Delta}\right)\Big|_{-\infty}^{+\infty} = 2\Delta C \,, \tag{4.67}$$

with

$$C = \frac{K_h}{2}\left(1 - \sqrt{1 - \frac{\dot{q}^2 M_S^2}{K_h^2 \gamma^2 \Delta^2}}\right) . \tag{4.68}$$

If we assume that we have just a small velocity $v = \dot{q}$, then we can make a Taylor expansion up to the second order in $\dot{q}$. Therefore, the energy becomes:

$$E_{K_h} \approx \frac{1}{2}\frac{M_S^2}{K_h \gamma^2 \Delta}\dot{q}^2 = \frac{1}{2}m\dot{q}^2 \, . \tag{4.69}$$

Here, we have defined the domain wall mass as:

$$m = \frac{M_S^2}{K_h \gamma^2 \Delta} \, . \tag{4.70}$$

Eq. (4.70) shows that the domain wall mass is connected with the hard axis anisotropy K_h. An increasing anisotropy K_h leads to an decrease of the domain wall mass. This also means that from a relativistic point of view the domain wall becomes faster with increasing K_h. We will see this behavior in the next section where we compare the analytical results with computer simulations.

Now the question appears what happens if $K_h \to 0$. In this case the domain wall mass diverges. In the limit $K_h = 0$ the mass is infinity following the definition Eq. (4.70). It seems that in this case the given definition does not hold anymore. An infinite mass means that the domain wall will not move. However, we have calculated the velocity of a domain wall with $K_h = 0$. It seems that in this case it is not possible to speak about a domain wall with a mass anymore. Therefore, M. Yan has proposed to call these kind of domain walls massless [31].

4.3 Numerical investigation of field driven domain walls

In this section we will prove the analytical predictions via numerical simulations. Here I will restrict myself first to the field driven domain wall motion. The current driven dynamics will be described separately after this section.

In the following, we investigate a transverse domain wall in linear magnetic chains oriented along the z-direction. Such chains can be interpreted as quasi 1D magnets [32, 33, 34] or as a cylindrical nanowire with a transverse domain wall [27, 28, 35, 36]. In the second case each spin has to be interpreted as the magnetization of one layer in the xy-plane $M_\alpha = \frac{1}{N}\sum_{n=1}^{N} S_n^\alpha$, $\alpha \in \{x,y,z\}$, where N is the number of spins in the layer. In such a system and in the case of the transverse domain wall the dipolar coupling plays an important role but leads to an increase of the easy axis anisotropy $D_e = D_u + D_d$ only. Here D_u is the uniaxial anisotropy coming from the spin-orbit coupling and the second term $D_d \approx 3\zeta(3)\mu_0\mu_s^2/(4\pi a^3)$ describes the shape anisotropy coming from the dipolar interaction [27] with $\zeta(3) \approx 1.202$.

The magnetic properties are described by the following biaxial Heisenberg Hamiltonian

$$\mathcal{H} = - J\sum_n \mathbf{S}_n \cdot \mathbf{S}_{n+1} - \mu_s \mathbf{B} \cdot \sum_n \mathbf{S}_n \\ + D_h \sum_n (S_n^x)^2 - D_e \sum_n (S_n^z)^2 , \qquad (4.71)$$

where the $\mathbf{S}_n = \boldsymbol{\mu}_n/\mu_s$ are three dimensional magnetic moments of unit length on a simple cubic lattice with the lattice constant a.

The first term in Eq. (4.71) describes the exchange coupling between nearest neighbors with ferromagnetic coupling constant $J > 0$. The second sum is the so-called Zeeman term, which describes the coupling of the spins to an external magnetic field $\mathbf{B}$. The third sum describes an easy zy-plane or hard x-axis anisotropy ($D_h > 0$). The fourth sum gives an easy axis anisotropy ($D_e > 0$) in z-direction which breaks the easy plane symmetry.

The underlying equation of motion which describes the dynamics is the Landau-Lifshitz-Gilbert (LLG) equation,

$$\dot{\mathbf{S}}_n = -\frac{\gamma}{(1+\alpha^2)\mu_s}\mathbf{S}_n \times \mathbf{H}_n + \frac{\alpha\gamma}{(1+\alpha^2)\mu_s}\mathbf{S}_n \times (\mathbf{S}_n \times \mathbf{H}_n) \qquad (4.72)$$

with the gyromagnetic ratio γ, the dimensionless Gilbert damping α, and the internal field given by the gradient $\mathbf{H}_n = -\partial \mathcal{H}/\partial \mathbf{S}_n$. The first term of the LLG equation describes the precessional motion of $\mathbf{S}_n$ and the second term the relaxation.

4 Field and current driven domain wall motion

Transverse domain wall motion

The simulations start with a relaxed head to head transverse domain wall, where the spins in the domains are oriented in $\pm z$-direction and inside the domain wall in y-direction. After switching on the external field the velocity of the domain wall can be calculated either by measuring the wall displacement (zero-crossing of the z-component of the magnetization) or from the time-dependence of the magnetization as $v = \frac{1}{2}\mathrm{d}\left(\sum_n S_n^z\right)/\mathrm{d}t$. Both methods lead to the same results.

Depending on the anisotropy one finds different cases of the magnetization reversal in combination with different velocities. Fig. 4.3 shows the velocity as a function of the Gilbert damping α [Fig. 4.3(a)] and external field $\mu_s B/J$ [Fig. 4.3(b)] for different hard axis anisotropies D_h/J and a constant easy axis anisotropy $D_e/J = 0.01$.

According to the literature the assumption of a zero hard axis anisotropy $D_h/J = 0$ leads to the precessional domain wall motion during magnetization reversal [27, 28, 35, 36]. We find this behavior in our simulations as well [see Fig. 4.3(a)]. The corresponding velocity equation

$$v_S = \frac{\gamma B}{\alpha + \frac{1}{\alpha}} \sqrt{\frac{Ja^2}{2D_e}}, \tag{4.73}$$

for a constant domain wall width $\Delta = \sqrt{Ja^2/(2D_e)}$ and $1/(\alpha + 1/\alpha)$ velocity dependence has been derived by J. Slonczewski [29] and confirmed numerically [27]. In the limit of zero damping ($\alpha = 0$) the velocity is zero and the magnetic moments inside the domain wall just precess around the z-axis. This behavior becomes obvious when one bears in mind that for $\alpha = 0$ the relaxation term in the LLG equation becomes zero.

If the hard axis anisotropy is finite the regime which was first described by L. R. Walker [37, 38] is reached [see Fig. 4.3(b)]. Here one finds a direct spin flip reversal of the magnetization during the domain wall motion which corresponds to a $1/\alpha$ behavior of the velocity:

$$v_W = \frac{\gamma B}{\alpha} \sqrt{\frac{Ja^2}{2\left(D_e + D_h \sin^2 \phi\right)}}. \tag{4.74}$$

Fig. 4.4 shows the dynamics of a transverse domain wall without hard axis anisotropy $D_h = 0$ [Fig. 4.4(a)] and with hard axis anisotropy $D_h \neq 0$ [Fig. 4.4(b)]. The figure clearly shows that the dynamics strongly depends on the hard axis anisotropy: $D_h = 0$: precessional motion, $D_h \neq 0$: direct reversal. The dynamics can be described by the angle ϕ.

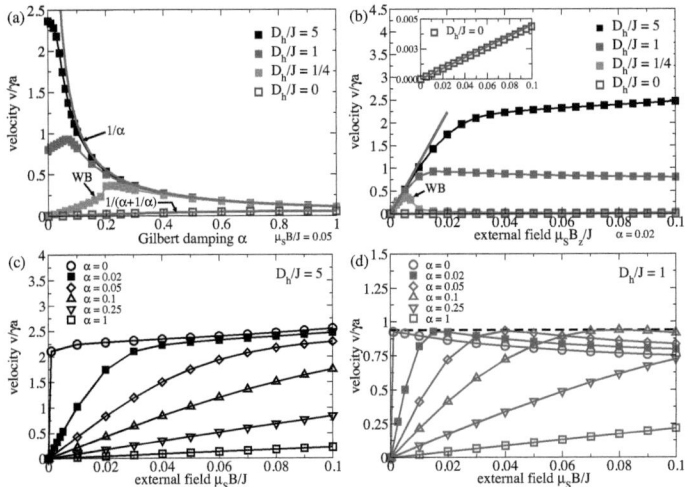

Figure 4.3: Velocity of a transverse domain wall with biaxial anisotropy (D_e and D_h) vs. Gilbert damping α (a), respectively external field $\mu_s B/J$ (b) - (d). WB marks the Walker breakdown and the solid lines in (a) and (b) represents the highest possible velocity given by Eq. (4.76). $D_e/J = 0.1$, $\mu_s B/J = 0.05$ [only (a)], $\alpha = 0.02$ [only (b)] are assumed.

The angle ϕ describes the angle between the plane where the spin motion takes place and the easy plane given by the hard axis anisotropy D_h. ϕ is constant in time but depends on the external field B, the hard axis anisotropy D_h, and the Gilbert damping α:

$$\phi = \frac{1}{2}\arcsin\left(\frac{\mu_s B}{\alpha D_h}\right). \tag{4.75}$$

In the limit of infinite hard axis anisotropy $D_h/J \to \infty$ the angle ϕ becomes zero which corresponds to a magnetization reversal strictly in the easy axis (xy)-plane. In this limit the Eq. (4.74) becomes

$$v_{\text{LL}} = \frac{\gamma B}{\alpha}\sqrt{\frac{Ja^2}{2D_e}}. \tag{4.76}$$

This equation has been derived by L. D. Landau and E. M. Lifshitz in 1935 [39] and gives the highest possible velocity of the domain wall.

4 Field and current driven domain wall motion

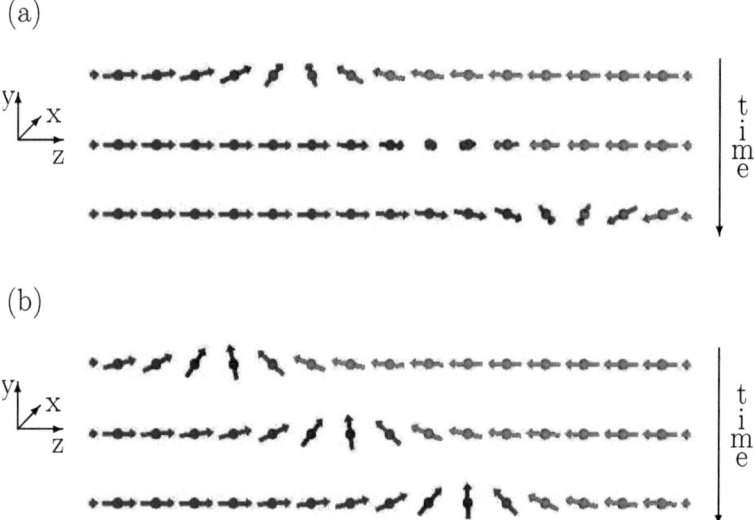

Figure 4.4: Dynamics of a transverse domain wall: (a) precessional motion without hard axis anisotropy D_h, (b) direct reversal $D_h \neq 0$. The figure shows a moving 1D transverse domain wall at three equidistant times.

4.3 Numerical investigation of field driven domain walls

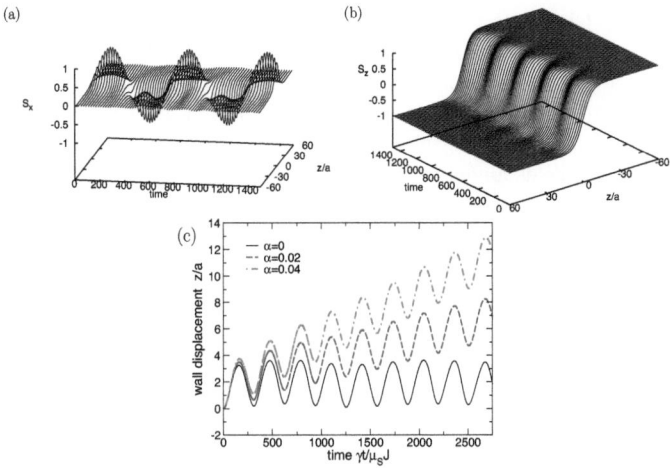

Figure 4.5: Walker breakdown: (a) and (b) S_x and S_z as function of time t and lattice site z, (c) domain wall position as function of time t.

Apart from the highest velocity Eq. (4.75) also defines the condition for the validity of the Walker formula. For a constant Gilbert damping α, a maximum value $B_{\max}$ exists (or vice versa a minimum value of the Gilbert damping $\alpha_{\min}$ exists for a constant external field B) beyond which the Eq. (4.75) is no longer fulfilled. This means that the spin motion is no longer restricted to a plane and an irregular precessional motion appears. This irregular precession in combination with an asymmetric oscillating force leads to a periodic change of the domain wall width and position [37, 40] which corresponds to the Walker breakdown (see Fig. 4.5) and leads to a decrease of the velocity as one could see in Fig. 4.3(a) and (b) for $D_h/J = 0.25$. In the limit of zero damping ($\alpha = 0$) one finds a periodic force and back motion of the domain wall with no effective wall displacement, which means that the time averaged velocity $\langle v \rangle$ vanishes.

A further increase of the hard axis anisotropy leads to the elimination of the Walker breakdown. Here one finds a finite velocity of the domain wall already for $\alpha = 0$ [see the $D_h/J = 1$ and $D_h/J = 5$ curves in Fig. 4.3(a)]. This is still true for an infinite hard axis anisotropy. Here the velocity is well described by the formula of Walker (Eq. 4.74) up to a maximum field value [Fig. 4.3(b)] or down to a minimum Gilbert damping α [Fig. 4.3(a)]. Beyond these limiting values the domain wall starts to emit spin waves. One can recognize that by comparing the curves for $D_h/J = 1$ and $D_h/J = 5$ with their $D_h/J = 0.25$ counterpart in Fig. 4.3(a) and (b). In contrast to $v(\alpha)$ for $D_h/J = 0.25$ the $v(\alpha)$ dependencies for stronger anisotropies ($D_h/J = 1$ and $D_h/J = 5$) significantly deviate from the $1/\alpha$ behavior but the velocity does not vanish for vanishing α [see Fig. 4.3(a)]. Additionally, the domain wall velocity for a fixed α value does not show an abrupt decrease for high fields anymore [see Fig. 4.3(b)] .

Fig. 4.3(c) and (d) give a more detailed analysis of the velocity as function of the external field for $D_h/J = 5$ and $D_h/J = 1$, respectively. Here, several α values are inspected. As expected an increase of the Gilbert damping leads to a suppression of the spin wave [29]. This corresponds to a linear increase of the velocity with increasing magnetic field. With decreasing damping the velocity increases and one can distinguish between two regions. The (nearly) linear increase of the velocity with increasing field corresponds to a domain wall motion without spin waves, while the velocity saturation corresponds to a motion with a spin wave wake. Further, the comparison between Fig. 4.3(c) and (d) shows that although the Walker breakdown has been eliminated the velocity still decreases after a maximum velocity [marked by the dashed line in Fig. 4.3(d)] has been reached. This behavior shows that there is a continuous crossover from Walker breakdown to a spin wave damped motion, where the velocity also decreases after reaching its maximum value. With decreasing hard axis anisotropy D_h the maximum velocity as well as the corresponding field value $B_{\max}$ decrease. In the case of Fig. 4.3(c) the turning point appears at high fields outside the plot.

With this information at hand it is possible to understand the behavior in the region of low Gilbert damping α. The $D_h/J = 1$ curve in Fig. 4.3(a) has been calculated with a field value $\mu_s B/J = 0.05$ which leads to the decreasing velocity with decreasing damping. The same dependence calculated with a driving field of $\mu_s B/J = 0.005$ would lead to the same behavior as in the case of $D_h/J = 5$, which means that the domain wall emits less spin waves and moves with a higher velocity.

Domain wall deceleration by spin waves

To understand the physics of domain wall deceleration by spin waves described in the last section one has to answer the following two questions: What is the driving mechanism and is it possible to give a quantitative description of the spin wave emission? To answer these questions we repeat all calculations of section 4.3 in the zero damping limit ($\alpha = 0$) in order to neglect the effect of all damping mechanism which are included in the Gilbert damping term [41, 21]. When the driving field is switched on the domain wall starts to move. During the period between starting time t_0 and a certain time t_e the domain wall relaxes. In this starting phase

4.3 Numerical investigation of field driven domain walls

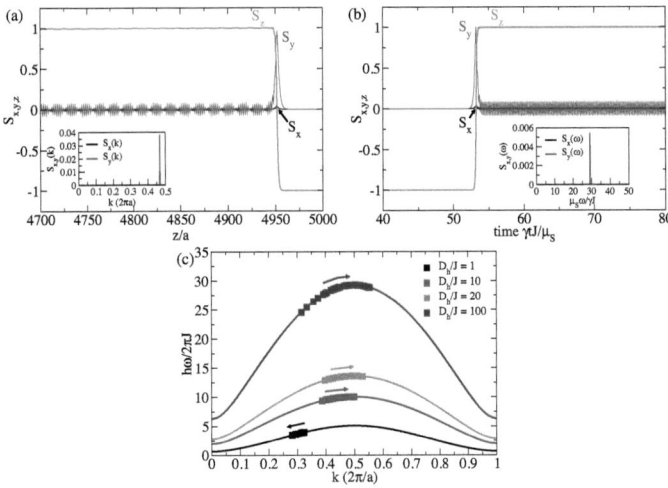

Figure 4.6: Spin wave behind a moving transverse domain wall: (a) magnetization as function of space (fixed time), (b) magnetization as function of time (fixed space). The insets show the corresponding Fourier transformations in k-, respectively ω-space. (c) Magnetic spin wave dispersion. Solid lines are the analytical solutions given by Eq. (4.77). The arrows mark the direction of change (increasing / decreasing) of the wave number k with increasing applied field $\mu_s B/J$. $D_e/J = 0.01$, $D_h/J = 100$ [only (a) and (b)], $\alpha = 0$ and $\mu_s B/J = 0.005$ [only (a) and (b)], $\mu_s B/J$ between $1 \cdot 10^{-6}$ and 1 [only (c)] are assumed.

the angle ϕ (see Eq. 4.75) changes and the velocity of the domain wall increases. During the relaxation the Zeeman energy is absorbed by the domain wall, after t_e no additional energy could be stored inside the domain wall, and the domain wall starts to emit spin waves.

With this information the driving mechanism for the spin wave emission becomes clear. The spin wave emission is the same as the wake behind a motorboat. This means that the spin wave is excited by the magnetization reversal of the spins inside the domain wall. The energy comes from the external field. The unexpected result of our calculations is that the domain wall moves at a constant magnetic field despite the fact that the relaxation term in the LLG equation [Eq. (4.72)] is skipped if $\alpha = 0$. This effect can be understood taking into account the shape of the precessional motion of magnetic moments. For circular precession no domain

4 Field and current driven domain wall motion

wall motion could occur. This is the case for $D_h/J = 0$. In the general case, however, the hard axis anisotropy distorts the shape of the orbit and makes it elliptic. The longer axis of the ellipse spans between the two minima of the easy axis anisotropy. That's why the switching between these two minima becomes possible. The exchange interaction ensures an energy transfer to the neighboring spins [42], i.e. makes the emission of spin waves behind the domain wall possible. The magnetic moments remain reversed.

For a quantitative description of the spin wave spectra the distance dependence of the spin wave at a fixed moment of time [Fig. 4.6(a)] and the time dependence at a fixed lattice point [Fig. 4.6(b)] have to be analyzed. For that purpose a Fourier transformation in space [see inset of Fig. 4.6(a)] and time [see inset of Fig. 4.6(b)] has to be performed [43, 44]. In both cases distinct peaks in the Fourier spectra have been found. Using these peaks the spin wave dispersion can be plotted. The symbols in Fig. 4.6(c) are the numerical data for different hard axis anisotropies D_h and field values between $\mu_s B/J = 1 \cdot 10^{-6}$ and $\mu_s B/J = 1$. Each point corresponds to a single field value. The solid lines are the corresponding analytical curves given by the formula (the derivation is presented in the Appendix):

$$\frac{\hbar\omega}{J} = 2S\sqrt{\left(1 - \cos(ka) + \frac{D_e}{J}\right)\left(1 - \cos(ka) + \frac{D_e}{J} + \frac{D_h}{J}\right)}. \qquad (4.77)$$

Next, we would like to compare these analytical expressions with the numerical data. The equation (4.77) has been derived under the assumption of zero external field **B**. In our simulations, however, one needs to apply a finite field in order to get domain wall motion. Therefore, each symbol in Fig. 4.6(c) corresponds to a certain field value $\mu_s B/J$.

The interesting point of the numerical solution is that the k-value has a non-trivial dependence on B. It increases with increasing field for higher values of D_h [the three upper curves in Fig. 4.6(c)], while it decreases for weaker D_h [bottom curve in Fig. 4.6(c)]. This is a direct consequence of the increasing / decreasing velocity with increasing field described above [see Fig. 4.3(c) and (d)]. In other words the wave number k corresponds to the energy gained during the magnetization reversal which increases (decreases) with increasing (decreasing) domain wall velocity [29].

With this information the physics becomes quite simple. After reaching the equilibrium the domain wall starts to emit spin waves of a certain wave number k and frequency ω. Because $\alpha = 0$ has been chosen the whole energy of the system is given by the sum of the domain wall energy and the energy of the spin waves. In equilibrium the energy of the domain wall reaches some constant value depending on the in-plane angle ϕ, the applied field B, and the anisotropy D_h.

4.3 Numerical investigation of field driven domain walls

The magnetization reversal, or in other words the moving domain wall, excites spin waves of a given constant energy consisting of a kinetic $\hbar\omega$ and a potential energy contribution coming from the Zeeman term. The potential part is defined via the canting angle θ formed between the magnetic moment and the z-axis while the kinetic energy is mostly determined by D_h and B.

Both energy contributions strongly depend on the reversal mechanism. In most cases the reversal is quite complicated and a mixture between the precessional motion and the direct reversal occurs. That's why the velocities found in our simulations lie somewhere between the highest possible velocity given by the formula of Landau and Lifshitz (Eq. 4.76) and the lower limit given by the formula of Slonczewski (Eq. 4.73). This also means that it is impossible or quite difficult to give an analytical prediction for the velocity or the functional dependence between field values $\mu_s B/J$ and the wave number k or the potential energy.

Connection to Solitons

While spin wave emission behind a moving domain wall is an effect almost unknown in the magnetism community a similar phenomenon is well known in the theory of solitons.

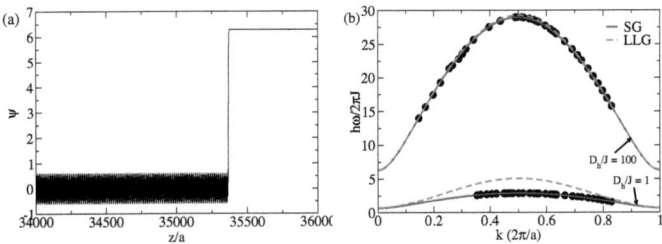

Figure 4.7: Waves behind a magnetic kink ψ as function of space (a) and spin wave dispersion (b). SG describes the solitonic dispersion given by Eq. (4.80) and LLG the magnetic dispersion [Eq. (4.77)]. The assumed values are: $D_e/J = 0.01$, $D_h/J = 100$ [only (a)], $\Gamma = 0$ and $f = 0.005$ [only (a)], $f = 7.5 \cdot 10^{-7} \ldots 0.4$ (b).

Following H. J. Mikeska [45, 32, 46, 47] one can map the LLG equation [Eq. (4.72)] to the Sine-Gordon equation. The necessary restriction is the ex-

4 Field and current driven domain wall motion

istence of the easy plane geometry. The procedure leads, then, to a damped double Sine-Gordon equation:

$$\frac{\partial^2 \psi}{\partial t^2} - c_0^2 \frac{\partial^2 \psi}{\partial z^2} + \omega_0^2 \sin\psi + f \sin\frac{\psi}{2} + \Gamma\frac{\partial \psi}{\partial t} = 0 \ . \tag{4.78}$$

Here $\psi = 2\theta$ with θ describing the profile of the domain wall: $\cos(\theta) = \pm\tanh(z/\Delta)$, where Δ is the domain wall width; $c_0 = \sqrt{\frac{2JD_h\gamma^2}{a\mu_s}}$; $\omega_0 = \sqrt{\frac{4D_eD_h\gamma^2}{a^3\mu_s}}$; $f = \frac{2\gamma^2 D_h B}{\mu_s}$ is the driving field, and $\Gamma = \frac{\gamma D_h\alpha}{\mu_s}$ the damping constant.

Using Eq. (4.78), we have performed the same simulation as before for the case of the magnetic domain walls. Starting with the relaxed static Kink solution ($f = 0$, $\Gamma = 0$) of Eq. (4.78) one gets:

$$\psi = 4\arctan[\exp(\pm z/\Delta)] \ . \tag{4.79}$$

A driving field $f \neq 0$ accelerates the Kink. After reaching the equilibrium the Soliton starts to emit spin waves. Then, the spin wave dispersion for $\Gamma = 0$ can be calculated in the same manner as before, with aid of Fourier transformations in space and time. Fig. 4.7(b) shows the comparison between the magnetic spin wave dispersions (LLG) given by Eq. (4.77) and the Sine-Gordon spin wave dispersions (SG) given by [47]

$$\omega^2 = 2c_0^2[1 - \cos(ka)]/a^2 + \omega_0^2 \ . \tag{4.80}$$

As one can see there is an excellent agreement between the numerical data (points) and the analytical curves (solid lines) given by Eq. (4.80). The dashed lines represent the analytical dispersion curves given by Eq. (4.77). It can be seen that for small D_h the magnetic spin wave (LLG) and Sine-Gordon (SG) solutions differ, while with increasing hard axis anisotropy D_h, the dispersion curves of the magnetic and soliton system become identical. This can be understood on the basis of the fact that with increasing anisotropy D_h the reversal becomes more direct ($\dot{\phi} \to 0$) and the magnetization reversal takes place in the easy plane given by D_h. In this case the angle ϕ [see Eq. (4.75)] becomes zero. This assumption, however, just leads to the double Sine-Gordon equation. For $D_h/J = 100$ this assumption is nearly fulfilled and both dispersion curves (magnetic and solitonic) coincide. This is not the case for $D_h/J = 1$. Here, the assumption is not fulfilled and a strong deviation occurs. Hence, applications of the double Sine-Gordon equation are reasonable only for $D_h/J \gg 1$.

4.4 Combined spin and molecular dynamics study of field driven domain wall motion

Combined spin and molecular dynamics simulation

The goal of the previous subsection was to clarify the mechanism behind the damping process in the case of a field driven transverse domain wall. The underlying physics of the damping is a highly interesting topic which is still not well understood yet. The reason is the Gilbert damping is phenomenological which means that the most spin dynamics simulations do not address the physics (microscopic description) of the energy dissipation.

H. Suhl [41] was the first who was dealing with this topic. He has worked out the energy dissipation between spin system and lattice. However, it is an accepted fact that at least three systems are necessary to describe the damping. It is an unquestioned fact that the spin is nothing else than the angular momentum of the electron. Therefore, we have to take into account also the electrons. The electrons itself are part of an atom and interact with the atomic cores. Here, we can exclude excitations of the core itself, but due to the interaction, the core can change the position which leads to an excitation of the lattice (phonons). Now, we have to distinguish between an indirect spin-lattice interaction mediated by the electron and a direct one. In the case of the indirect spin-lattice interaction the spin excites the electron and the excited electron excites the lattice. In the case of the direct spin-lattice interaction the energy of the spin directly dissipates into the lattice, again mediated by the electron, but in this case without excitation of the electron. Fig. 4.8 shows the connection between the three systems: spin, electron, and lattice, which have to be taken into account for a complete description of the damping process in magnetism.

So far the description was justified to the spin system only. This is legitim as long as the energy dissipation has been taken into account by the Gilbert damping. Without such a damping, the domain wall starts to emit spin waves. The restriction to $\alpha = 0$ was the first step towards the understanding of the damping process. This step was possible because the damping constants α are quite small in real systems. During the next steps we have to take into account also the other two systems, electrons and lattice, to get a complete picture.

The description of the electrons is quite complex and should be part of a future investigation. The direct coupling between spin and lattice is not so difficult and shall be considered in the following. Until now, I have assumed that the lattice is fixed (Born-Oppenheimer approximation).[2] This means the position of the spins

---

[2]Born-Oppenheimer approximation: the mass of the core is 100 times larger than the mass of the electrons. Therefore, we can approximately say that the cores do not

4 Field and current driven domain wall motion

is constant in time. The spin itself can change the spacial direction but not its position in space. This assumption shall be skipped in the following. In the following, we allow the atoms to move.

Then, we have to write down the corresponding differential equations. We know, that the spin dynamics is well described by the Landau-Lifshitz-Gilbert equation. We further know, that the lattice can be described by molecular dynamics. Therefore, we have to combine spin and molecular dynamics by solving the Landau-Lifshitz-Gilbert equation together with, e.g., the Hamilton equations:

$$\dot{\mathbf{r}}_n = \frac{\partial \mathcal{H}}{\partial \mathbf{p}_n} \quad \text{and} \quad \dot{\mathbf{p}}_n = -\frac{\partial \mathcal{H}}{\partial \mathbf{r}_n} , \tag{4.81}$$

or written as Newton's differential equations:

$$\dot{\mathbf{r}}_n = \mathbf{v}_n \quad \text{and} \quad \dot{\mathbf{v}}_n = \frac{\mathbf{F}_n}{m} . \tag{4.82}$$

Here, $\mathbf{r}_n$ is the position, $\mathbf{p}_n$ is the momentum, and $\mathbf{v}_n$ the velocity of the nth atom, m is the mass of the atoms, and $\mathcal{H}$ the Hamilton function. The force $\mathbf{F}_n$ is given by:

$$\mathbf{F}_n = -\frac{\partial \mathcal{H}}{\partial \mathbf{r}_n} + \mathbf{F}_{\text{ext}} , \tag{4.83}$$

where $\mathbf{F}_{\text{ext}}$ are external and $-\partial \mathcal{H}/\partial \mathbf{r}_n$ the internal forces.

Now, we have to think about the Hamilton function $\mathcal{H}$ itself. The Hamilton function has to include terms describing the lattice as well as the spin system. For the spin system we can use the known Heisenberg Hamiltonain Eq. (4.71). Furthermore, we have to make sure that we couple spin and lattice. A Hamilton function which fulfills all these conditions is given by:

$$\mathcal{H} = -\sum_{\langle ij \rangle} J(\mathbf{r}_{ij}) \mathbf{S}_i \cdot \mathbf{S}_j + \mathcal{H}_{\text{mag}}(\mathbf{S}_i) - \sum_i \frac{\mathbf{p}_i^2}{2m} + \frac{C}{2} \sum_{\langle ij \rangle} (\mathbf{r}_i - \mathbf{r}_j)^2 . \tag{4.84}$$

The first term describes the exchange interaction which is depending on the distance between the spins. In the simplest case, $J(\mathbf{r}_i)$ is given by:

$$J(\mathbf{r}_{ij}) = J_0 \exp\left[-\kappa \left(|\mathbf{r}_i - \mathbf{r}_j| - a\right)\right] . \tag{4.85}$$

$J(\mathbf{r}_{ij})$ describes an exponential decay of the exchange interaction with the distance $|\mathbf{r}_i - \mathbf{r}_j|$. This term describes the interaction between nearest neighbor spins. The next nearest (nn) and next next nearest (nnn) neighbor exchange interactions shall

---

change their positions with time.

4.4 Combined spin and molecular dynamics study of field driven domain wall motion

be negligible. The same is true for all longer distances. κ is the decay constant, which shall fulfill this condition, $\mathbf{r}_i$ and $\mathbf{r}_j$ are the positions of spin i and j, and a is the lattice constant. If the atoms are in rest (no lattice distortion) $a = |\mathbf{r}_i - \mathbf{r}_j|$, and therefore $J(\mathbf{r}_{ij})$ becomes: $J(\mathbf{r}_{ij}) = J_0$.

The second term describes all magnetic interactions which do not depend on $\mathbf{r}$, like the Zeeman term and the uniaxial anisotropies (easy and hard axis anisotropy). The third term describes the kinetic E_{kin} and the last term the potential energy E_{pot} of the atoms. For simplicity an harmonic oscillator potential has been assumed as lattice potential. This is an approximation which is based on the assumption that we have just a small lattice distortion. The same assumption can be found in any textbook describing phonons. In the case of larger lattice distortions we have to replace the harmonic oscillator potential by a more realistic potential like the Morse potential. Nevertheless, the description does not change in these cases.

The assumption of small lattice distortions makes it possible to expand the exponential function in the exchange interaction in a series. Here, we take into account only the offset and the linear term:

$$J(\mathbf{r}_{ij}) \approx J_0 \left[1 - \kappa \left(|\mathbf{r}_i - \mathbf{r}_j|\right)\right] . \tag{4.86}$$

Therefore, we can write the Hamilton function Eq. (4.84) as:

$$\mathcal{H} = -J_0 \sum_{\langle ij \rangle} \mathbf{S}_i \cdot \mathbf{S}_j - \sum_{\langle ij \rangle} \kappa_{ij} \mathbf{S}_i \cdot \mathbf{S}_j + \mathcal{H}_{\text{mag}}(\mathbf{S}_i) + \sum_i \frac{\mathbf{p}_i^2}{2m} + \frac{C}{2} \sum_{\langle ij \rangle} (\mathbf{r}_i - \mathbf{r}_j)^2 , \tag{4.87}$$

with

$$\kappa_{ij} = J_0 \kappa \left(|\mathbf{r}_i - \mathbf{r}_j| - a\right) . \tag{4.88}$$

The first term is now the well known nearest neighbor exchange interaction which is independent of the distance between the two spins $\mathbf{S}_i$ and $\mathbf{S}_j$. The new second term describes the coupling between spin and lattice, with the coupling constant $\kappa_{ij} \propto J_0 \kappa$. The other terms are identical to the previous Hamilton function Eq. (4.84). This new Hamiltonian represents the way of a combined spin and molecular dynamic simulation. It contains terms describing the spin system, terms describing the lattice, and a coupling between spin and lattice.

Simulation results

The simulation itself is similar to the simulation described in the previous section. However, instead of the Heisenberg Hamiltonian Eq. (4.71) the Hamiltonian Eq. (4.84) has been used.

4 Field and current driven domain wall motion

Fig. 4.9 shows the results of the simulation: 4.9 (a) shows the magnetization as function of lattice site x and 4.9 (b) as function of time t. These figures are similar to Fig. 4.6 (a) and (b) in the previous section. The only difference is, that the higher field value here leads to an additional oscillation of S_z. Fig. 4.9 (c) and (d) show the momentum p_z as function of the lattice site x and time t. As before the spins the lattice behind the domain wall show an oscillation. The spin oscillation shows the appearance of a spin wave. In the case of the lattice the oscillation is connected to phonons. The large peak signals the domain wall, respectively the time when the domain wall passes the measuring point.

To get some more information we have to overlay the spin and momentum signal. The result is shown in Fig. 4.10. The comparison of both oscillations shows that the spin oscillation (spin wave) is twice the frequency of the oscillation of the momentum. Further, the spin oscillation is an oscillation around zero while the momentum oscillates around a positive value. Therefore, it is not clear if we shall call this oscillation a phonon. Furthermore, for a phonon we have to look for the lattice distortion itself. The lattice distortion is shown in Fig. 4.11 together with the momentum oscillation. It can be clearly seen that there is a lattice distortion. However, it is not a fast oscillation as we expect for phonons. If there is a lattice oscillation (phonon), then it is on a longer time scale than the simulation time.

Some words about the momentum oscillation in Fig. 4.11: Fig. 4.9 shows the oscillation shortly behind the domain wall resp. close to it. If we zoom out as in Fig. 4.11 we see an exponential decreasing of the oscillation. The same can be seen in principle also for the spin waves. It just shows that the phonons, respectively the spin waves, move to the opposite direction compared to the domain wall. This is similar to the wake behind a boat.

4.4 Combined spin and molecular dynamics study of field driven domain wall motion

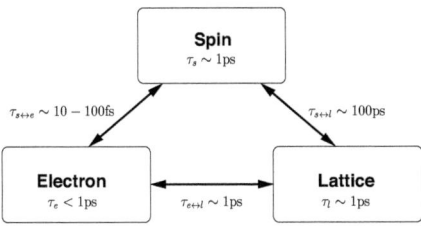

Figure 4.8: The dynamical behavior of a magnetic system can be understood in terms of three thermodynamic reservoirs and the interactions between these reservoirs. The given times are the lifetimes resp. the relaxation times.

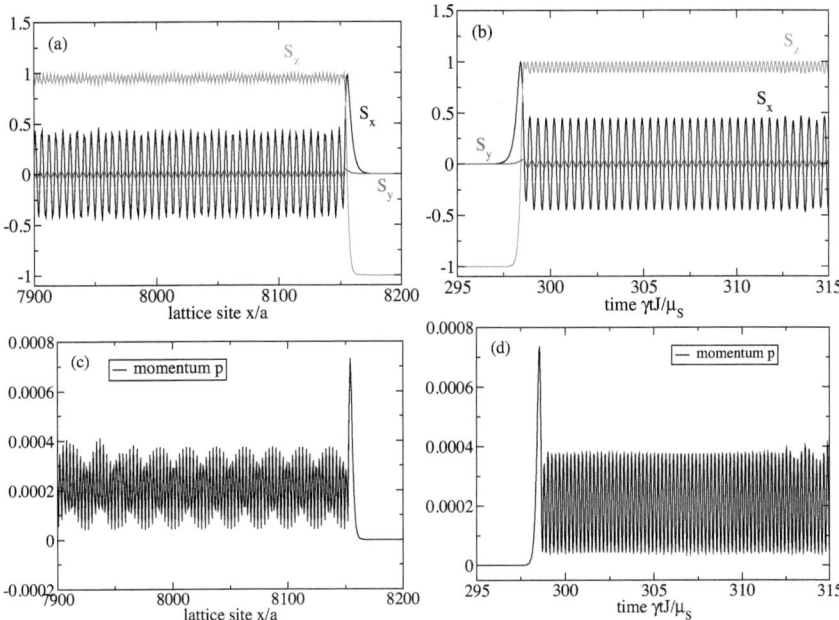

Figure 4.9: Spin waves and lattice momentum behind a field driven transverse domain wall: (a) and (b) show the magnetization as function of lattice site x (fixed time), and as function of time t (fixed place). (c) and (d) show the corresponding lattice momentum as function of lattice site x and time t.

69

4 Field and current driven domain wall motion

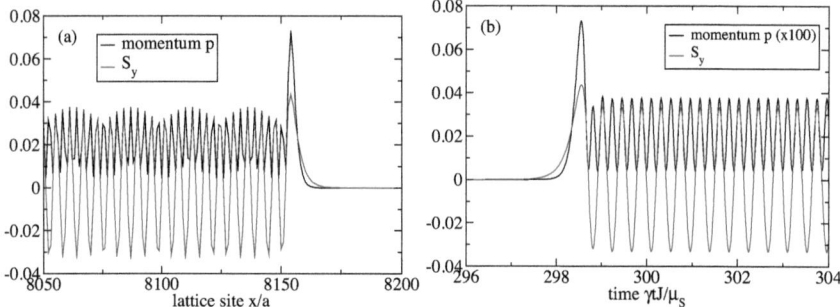

Figure 4.10: Comparison between the oscillations of the magnetization (spin waves) and the lattice momentum. Fig. (a) shows the oscillations as function of the lattice site (fixed time) and (b) as function of time at a certain lattice site.

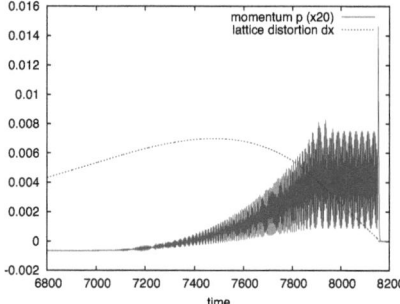

Figure 4.11: Oscillation of the lattice momentum and lattice distortion behind a moving transverse domain wall.

4.5 Numerical investigation of current driven domain walls

Current driven domain wall motion from a technological point of view

In Sec. 4.2 the velocity of a field and current driven transverse domain wall has been calculated analytically. The correctness of the results with respect to the field driven domain wall motion has been proven in the previous section 4.3. In this section we will concentrate ourself on the current driven domain wall dynamics.

From a technological point of view the current driven domain wall motion is more important as the field driven domain wall dynamics. The reason therefore is first of all the controllability. Magnetic fields are harder to control than electric currents. The response time and the possibility to focus spatially are much better in the case of electric currents as in the case of magnetic fields. Furthermore, in most cases electric currents are used to produce magnetic fields. However, this also means power loss; the effectivity is not 100%. Therefore, it is clear that the direct use of an electric current is more effective.

Another reason is the fact that due to the conservation of the torque two current driven 180° transverse domain walls always move in the same direction without changing the distance between the domain walls. A magnetic field would move the domain walls in opposite directions. This means that the distance between the domain walls changes. The final result can be the annihilation of domain walls. However, if we want to use the domain walls or the domains in between for information storage magnetic fields are not the appropriate technology to drive the domain walls. S. Parkin has proposed to use thin magnetic wires with more than one domain wall to store magnetic information. The device is called magnetic racetrack memory [48] and needs spin-polarized currents to shift all domain walls in one direction.

Comparison with analytical results

As before in Sec. 4.3 we will prove the correctness of the analytical results with respect to the influence of the electric current and go beyond the analytical results. In the following the transverse domain walls are only driven by an electric current and not by a magnetic field. This means in the following we assume $B = 0$.

In the previous Sec. 4.2 we have seen that the domain wall dynamics strongly depends on the reversal mechanism and the reversal mechanism on the existence of a hard axis anisotropy D_h (K_h). If $D_h \neq 0$ (with hard axis anisotropy) the reversal will be direct and the domain wall motion will be straight. In the case of no hard axis anisotropy $D_h = 0$ the reversal is a precession which leads to a precessional motion of the transverse domain wall. In both cases we can expect different velocities of the domain wall.

4 Field and current driven domain wall motion

Let us start with $D_h = 0$. The results of these simulations are presented in Fig. 4.12 which shows the domain wall velocity v as function of the Gilbert damping α as well as the strength of the current u (inset). The simulations have been performed for different values of the non-adiabatic parameter β. In all cases

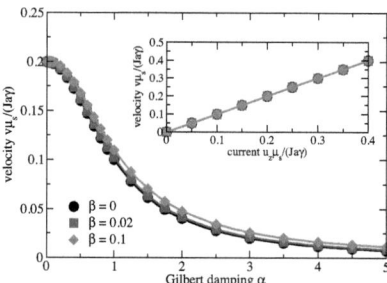

Figure 4.12: Domain wall velocity of a TDW as a function of the Gilbert damping α for different values of β in cylindrical wires. Here $u = 0.2$ is assumed. The inset shows the velocity as a function of u for $\alpha = 0.02$.

we find the expected velocitiy [Eq. (4.49) in Sec. 4.2]:

$$v = \frac{1 + \beta\alpha}{(1 + \alpha^2)} u \ . \tag{4.89}$$

Fig. 4.13 shows the time evolution of the out-of-plane component of the transverse domain wall. With other words, Fig. 4.13 shows that the transverse domain wall precesses during the motion if $D_h = 0$. It can be seen that the sense of rotation depends on the relation between the Gilbert damping α and the non-adiabaticity β. The precession can be clockwise ($\alpha > \beta$) or anti-clockwise ($\alpha < \beta$). Furthermore, there is no precession if $\alpha = \beta$. The sense of rotation and the speed are given by Eq. (4.53):

$$v_{\text{prec.}} = \frac{\alpha - \beta}{1 + \alpha^2} u \ . \tag{4.90}$$

This equation immediately shows that there is no precession if $\alpha = \beta$ and a different sense of rotation for $\alpha < \beta$ and $\alpha > \beta$.

4.5 Numerical investigation of current driven domain walls

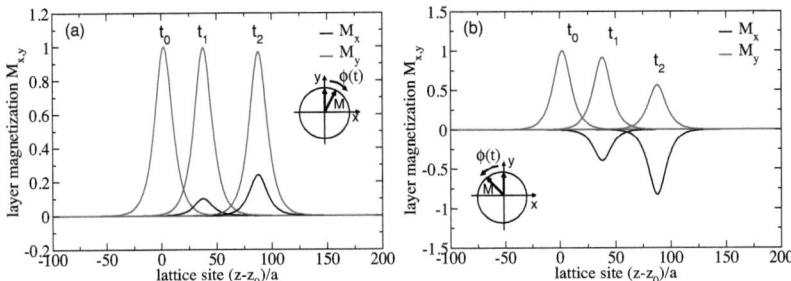

Figure 4.13: Time evolution of the out-of-plane components of a TDW corresponding to $u = 0.2$, $\alpha = 0.02$ and (a) $\beta = 0$, (b) $\beta = 0.1$; $t_0 < t_1 < t_2$ are equidistant steps in time. The wire crosssections are depicted in the insets.

Let us come now to the situation with hard axis anisotropy ($D_h \neq 0$). The hard axis anisotropy leads to a direct reversal and therefore to a straight motion. In the case of a field driven transverse domain wall with a hard axis anisotropy we have seen additional effects like the Walker breakdown and the emission of spin waves. The Walker breakdown has been predicted also in the case of current driven domain walls in Sec. 4.2. However, we will see that we can also expect the emission of spin waves in the case of current driven domain walls if the Walker breakdown is prevented. This is the case, e.g., for a huge hard axis anisotropy $D_h \gg 0$.

Fig. 4.14 shows the domain wall velocity v as function of the current u for different β values. As the analytical calculation predicts the velocity is given by Eq. (4.159):

$$v = \frac{\beta}{\alpha} u \,. \tag{4.91}$$

For $\beta = 0$ the Gilbert equation [Eq. (3.109)] contains only the adiabatic spin torque. In this case we expect from Eq. (4.91) that the velocity of the domain wall is zero. This can be seen also in Fig. 4.14(a). However, there is a critical current $u_{\text{crit.}}$ after which the domain wall starts to move: $u > u_{\text{crit.}} \Leftrightarrow v > 0$. The explanation for that is quite simple. We have to remember the scenario without hard axis anisotropy $D_h = 0$ and expect that for $\beta = 0$ the electric current drives the domain wall with a velocity given by:

$$v_{\beta=0, D_h=0} = \frac{\alpha}{1+\alpha^2} u \,. \tag{4.92}$$

4 Field and current driven domain wall motion

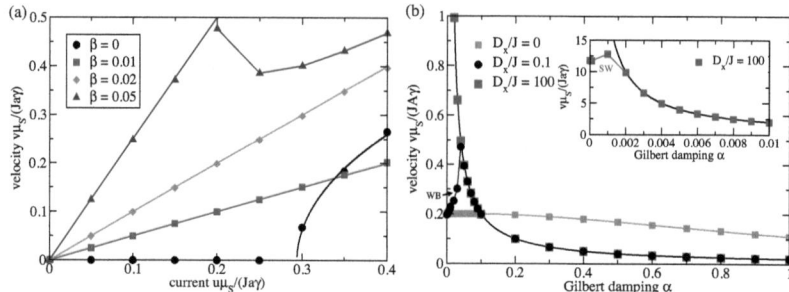

Figure 4.14: Velocity of a current driven transverse domain wall as function of the strenght of the driving current (a) and as function of the Gilbert damping α (b). Depending on the non-adiabaticity parameter β and on the strength of the hard axis anisotropy D_h, different reversal mechanism can be found leading to different velocities.

In this case the conservation of the spin torque (adiabatic spin torque) leads to a precessional motion which is not blocked due to the existence of a hard axis anisotropy. With hard axis anisotropy D_h the precession is forbidden as long as the current is not strong enough to push the spins in the direction of the hard axis anisotropy. This is the case for $u < u_{\text{crit.}}$. The non-adiabatic spin torque which is given by scattering processes of the free electrons with the magnetization of the domain wall and which could drive the domain wall has been explicitly skipped by the assumption $\beta = 0$. Therefore, the velocity is zero.

For $u > u_{\text{crit.}}$ the current forces the spins to overcome the energy barrier due to the hard axis anisotropy and the domain wall starts to move with a precessional motion similar to the scenario with $D_h = 0$. With other words, we have here a Walker breakdown. However, in this case the velocity increases after the Walker breakdown and does not decrease as the name Walker "breakdown" suggests. The critical current $u_{\text{crit.}}$ at which the Walker breakdown takes place depends on the strength of the hard axis anisotropy D_h and becomes zero for $D_h = 0$.

If we expect $\beta > 0$, we see a straight domain wall motion with a direct reversal. Furthermore, depending on the Gilbert damping α, the non-adiabaticity β, and the value of the hard axis anisotropy we find a critical current strength $u_{\text{crit.}}$ which we have to overcome ($u > u_{\text{crit.}}$) to see a real Walker breakdown where the nonzero velocity breaks down. In Fig. 4.14(a) for $\beta = 0.05$ such a Walker breakdown can be seen. For the other curves the Walker breakdown occurs at higher u. The description of the Walker breakdown is the same as before in the case of the field

4.5 Numerical investigation of current driven domain walls

driven domain wall motion or as for the scenario with $\beta = 0$ and $D_h \neq 0$. The spins inside the domain wall will be pushed by the current into the direction of the hard axis anisotropy. If the spin can overcome the energy barrier of the hard axis anisotropy the domain wall starts moving with a precessional motion and the velocity goes down. The stability condition is given by:

$$\phi = \frac{1}{2}\arcsin\left(\frac{(\beta - \alpha)\frac{uM_s}{\gamma\Delta}}{\alpha K_h}\right). \tag{4.93}$$

For the field driven transverse domain walls we have seen that the Walker breakdown can be prevented with a huge hard axis anisotropy. In these cases the domain wall starts to emit spin waves. In the following we will see that the same scenario appears also for the current driven domain walls. Fig. 4.14(b) shows the velocity v as function of the Gilbert damping α. The three curves correspond to $D_h/J = 0$, $D_h/J = 0.1$ and $D_h/J = 100$. In the case $D_h = 0$ we have a precessional motion and the velocity is given by Eq. (4.89). This means that for zero damping ($\alpha = 0$) the velocity is finite:

$$v_{\alpha=0} = u. \tag{4.94}$$

A small but nonzero hard axis anisotropy $D_h > 0$ means a direct reversal, but for $u > u_{\text{crit.}}$ or $\alpha < \alpha_{\text{crit.}}$ a Walker breakdown occurs. In these cases the velocity decreases and becomes equal to Eq. (4.94) in the limit $\alpha \to 0$. The decrease of the velocity is equal to the field driven domain wall motion. But, the final velocity is different. In the case of a field driven transvese domain wall with a precessional motion, the velocity in the limit $\alpha \to 0$ is zero. After the Walker breakdown the transverse domain wall shows a precessional motion and therefore the velocity becomes zero in this case for $\alpha \to 0$. However, in the case of a current driven domain wall we find a small but finite velocity after the Walker breakdown, even for $\alpha \to 0$.

If D_h is huge ($D_h \gg 0$) the velocity becomes infinite after Eq. (4.91). However, Fig. 4.14(b) (inset) shows that the velocity is huge and finite. This means that Eq. (4.91) does not predict the correct velocity in this limit.

The corresponding domain wall profile is shown in Fig. 4.15(a), while the corresponding time evolution of a single spin within the chain is presented in Fig. 4.15(b). Both figures show the appearance of a spin wave wake behind the domain wall. The spin waves can be analyzed by a Fourier transformation. The corresponding peaks are given as the insets in Fig. 4.15 and the result of the analysis is the dispersion relation shown in Fig. 4.16. The dots are the numerical results for three different values of u. In this case the energy decreases with increasing

75

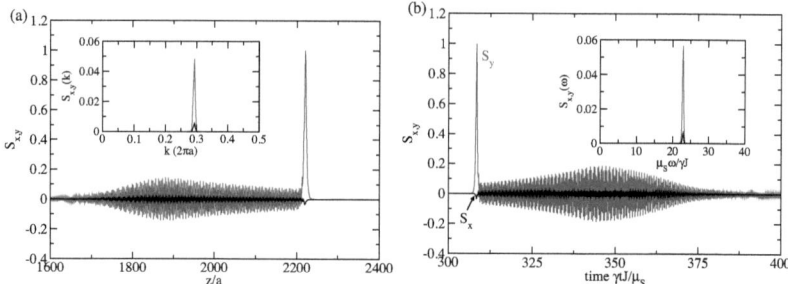

Figure 4.15: Spin wave wake behind a current driven transverse domain wall (no magnetic field). Fig. (a) shows the magnetization as function of the lattice side x at a fixed time and Fig. (b) shows the magnetization as function of time t at a fixed lattice side. The insets show the Fourier transformation of the spin waves behind the domain wall.

current marked by the arrow. The solid line is the analytical dispersion relation; it is the same as before for the field driven transverse domain walls:

$$\frac{\hbar\omega}{J} = 2S\sqrt{\left(1 - \cos(ka) + \frac{D_e}{J}\right)\left(1 - \cos(ka) + \frac{D_e}{J} + \frac{D_h}{J}\right)}. \quad (4.95)$$

Furthermore, the explanation is the same as before in Sec. 4.3. The domain wall excites the spin waves during the reversal process of the spins inside the domain wall during the motion. Due to $\alpha = 0$ (the same is true if α is very small $\alpha \ll 1$) the spins behind the domain wall continue precessing. Due to the huge hard axis anisotropy ($D_h/J = 100$) the reversal mechanism itself is a direct reversal even if the relaxation term in the Landau-Lifshitz-Gilbert equation is dislodged. Without hard axis anisotropy we would expect only a precessional motion and no relaxation in this case. Due to the fact that the hard axis anisotropy D_h forces a direct reversal we see a straight domain wall motion even for $\alpha = 0$. During this reversal we see a periodic change of the shape of the domain wall which leads to a periodic increase and decrease of the domain wall energy. The increase comes from the current which forces the magnetization into the hard axis direction. The energy decrease provides the energy for the spin waves. This decrease comes from the fact that the magnetization is not able to overcome the energy barrier due to the hard axis anisotropy and evades to reduce the domain wall energy. However, this leads to a periodic motion of the spins inside the domain wall which drives the domain wall forward and can be seen as continuing oscillation (spin wave) of

4.5 Numerical investigation of current driven domain walls

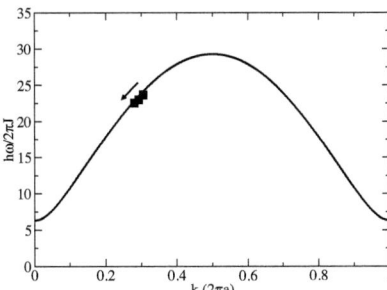

Figure 4.16: Energy dispersion of the spin wave behind the transverse domain wall. The points are the numerical results for three different current strength: $u_x = 0.1$, 0.2, and 0.3. The arrow shows the change of the energy corresponding to an increase of u_x. The solid line corresponds to the expected analytical energy dispersion Eq. (4.95).

the spins behind the domain wall.

This behavior is the same for both driving mechanisms: external field or current. However, in the case of the external field we could argue that the domain wall can only store a certain amount of energy and all the additional energy will be used to excite the spin waves. The energy itself comes from the external field (Zeeman term). With the domain wall motion more and more spins will be aligned in field direction. Therefore, the spin chain gains energy which will not disappear due to $\alpha = 0$ (no energy dissipation). In the case of a pure current driven domain wall motion we have no external field ($B = 0$) which leads to an increase of the system energy and therefore the previous argumentation cannot be used. However, the domain wall can be seen as a quasiparticle with a mass m (see Sec. 4.2) and a kinetic energy $E_{\text{kin.}} = 1/2 m v^2$ which is nonzero if the domain wall moves ($v \neq 0$). This means that the currrent pumps energy into the system. The spin torque is nothing else than a force acting on the spins and a force always corresponds to an energy. During the domain wall motion the spin torques due to the electric current change the shape and energy of the domain wall. More precise by the current leads to a change of the angle ϕ during the domain wall motion [see Eq. (4.158)] and this leads to a change of the domain wall shape and energy [see Eq. (4.8) and (4.28)]. If ϕ becomes larger than 90° the Walker breakdown sets in. This is not possible for a huge D_h. However, $\phi = 90°$ corresponds to the highest possible domain wall energy, it is not possible to store more energy in the domain wall. The additional energy will be used to excite the spin waves. If D_h is small (Walker breakdown) or zero this additional energy will lead to a precession of the domain wall. This is independent from the fact where the energy comes form: magnetic field or electric current.

Some additional remark

So far, we have discussed the possibility to describe the current driven domain wall motion analytically. Furthermore, we have seen that it is possible to simulate such processes. Both descriptions give a perfect agreement.

Now, we will consider the case when the transverse domain wall shows a precessional motion ($\dot{\phi} \neq 0$, no hard axis anisotropy $D_h = 0$) during movement. In this case it is not necessary to perform analytical or numerical calculations to get the velocity of a current driven domain wall. All the information can be obtained by analyzing the Landau-Lifshitz-Gilbert equation with the additional spin torque terms:

$$\begin{aligned}\frac{\partial \mathbf{S}_i}{\partial t} =& -\frac{\gamma}{(1+\alpha^2)\mu_s}\mathbf{S}_i \times \mathbf{H}_i - \frac{\alpha\gamma}{(1+\alpha^2)\mu_s}\mathbf{S}_i \times (\mathbf{S}_i \times \mathbf{H}_i) \\ &- \frac{\alpha-\beta}{(1+\alpha^2)}\mathbf{S}_i \times [(\mathbf{u}\cdot\nabla)\mathbf{S}_i] \\ &+ \frac{1+\alpha\beta}{(1+\alpha^2)}\mathbf{S}_i \times (\mathbf{S}_i \times [(\mathbf{u}\cdot\nabla)\mathbf{S}_i]) \,,\end{aligned} \qquad (4.96)$$

according to the Hamilton function in this special case:

$$\mathcal{H} = -J\sum_{\langle ij \rangle} \mathbf{S}_i \cdot \mathbf{S}_j - D_z \sum_i (S_i^z)^2 \,. \qquad (4.97)$$

A relaxed transverse domain wall leads to a "translational" and "rotational invariance" of the dynamics of such domain walls because $\mathbf{S}_i \times \mathbf{H}_i = 0$, i.e., the first and second term in Eq. (4.96), describing the precession and the relaxation due to the internal field $\mathbf{H}_i$, do not contribute to the time evolution of the magnetic moments. In this particular case the time evolution appears entirely due to the terms induced by the electric current, namely the third (relaxation) and fourth (precession) term in Eq. (4.96). If an additional hard axis anisotropy is present the situation changes: a rotation of the domain wall leads to a change of the internal field $\mathbf{H}_i$. Furthermore, depending on the sign of $\alpha - \beta$, the fourth term describes either a clockwise ($\alpha - \beta > 0$) or an anticlockwise ($\alpha - \beta < 0$) rotation. For $\beta = \alpha$ the precessional term is zero, and the time evolution is described solely by a direct reversal of the magnetization. In this particular case the velocity of the domain wall can be read off directly from the prefactor of the remaining third term, i.e.,

$$v_{\beta=\alpha} = u \,. \tag{4.98}$$

In the case $\alpha \neq \beta$, a similar derivation leads to the two decoupled terms on the right hand side of Eq. (4.96). The first term (initially the third term) is responsible for the precession. The second term (initially the fourth term) is responsible for the relaxation. Due to the above mentioned decoupling the velocity can be obtained directly from the prefactor of the relaxation term

$$v = \frac{1+\beta\alpha}{(1+\alpha^2)} u \,. \tag{4.99}$$

The precessional term in Eq. (4.96) leads to a decoupled rotation of the out-of-plane components of TDW's with a precession speed equal to

$$v_{\text{prec.}} = \frac{\alpha-\beta}{1+\alpha^2} u \,, \tag{4.100}$$

the sense of rotation as already mentioned being determined by the sign of $\alpha-\beta$. It is important to notice that these velocity equations (4.99) and (4.100) describe the motion of a current driven TDW performing a precessional motion. The results are independent of the wire geometry, similar results could be found also for thin film geometries, quadratic rods, or linear chains [49, 50, 51]. The only condition which has to be fulfilled is just a precession of the domain wall. That means that a current driven TDW after the Walker breakdown moves with a velocity which can be calculated using Eq. (4.99) (see e.q. [52, 53]).

4 Field and current driven domain wall motion

4.6 Influence of the DM interaction on the domain wall motion

In the previous sections we have discussed the dynamics of transverse domain walls in ferromagnetic structures. Such structures are called collinear. During the last years the Dzyaloshinsky-Moriya interaction (DMI) became more and more important. The DMI is described by the following Hamiltonian:

$$\mathcal{H}_{\mathrm{DM}} = \frac{\mathbf{D}_{\mathrm{DM}}}{2} \cdot \sum_{\langle ij \rangle} \mathbf{S}_i \times \mathbf{S}_j \qquad (4.101)$$

The DMI appears due to the symmetry breaking at the surface and can lead to non-collinear structures like spin-spirals. In these cases it is nearly impossible to define domain walls. If the DMI is weak the system stays collinear. However, the DMI influences the dynamics of the domain wall. The factor 1/2 has been introduced here to equalize the double summation.

This section shall clarify the nature and strength of the DMI. The section is based on two publications by Tretiakov et al. [54] and Thiaville et al. [55]. Both publications deal with transverse domain walls influenced by the Dzyaloshinsky-Moriya interaction. However, the assumptions are different: In the publication by Tretiakov et al. the vector $\mathbf{D}_{\mathrm{DM}}$ has been assumed to be parallel to the easy axis anisotropy $D_e(K_e)$ which is needed to get a transverse domain wall. Thiaville et al. have assumed a direct reversal of the domain wall which means the appearance of a hard axis anisotropy. Further, the vector $\mathbf{D}_{\mathrm{DM}}$ shall be parallel to the hard axis anisotropy $D_h(K_h)$ which shall be perpendicular to the easy axis anisotropy $D_e(K_e)$. Within this section we will deal with both scenarios. At first, we have to look for the micromagnetic description of the Dzyaloshinsky-Moriya interaction.

Micromagnetic description of the Dzyaloshinsky-Moriya interaction

The starting point is the Hamilton function (4.101). Then, the way to derive the DMI in the micromagnetic description is the same as before in chapter 2 for the exchange interaction. First, we have to replace the discrete spins $\mathbf{S}_i$ by the magnetization $\mathbf{S}(\mathbf{r})$ at the position $\mathbf{r}$:

$$\mathbf{S}_i \approx \mathbf{S}(\mathbf{r}), \qquad (4.102)$$

$$\mathbf{S}_{i\pm 1} \approx \mathbf{S}(\mathbf{r}) \pm a \left(\frac{\partial \mathbf{S}}{\partial \mathbf{r}} \right) \qquad (4.103)$$

4.6 Influence of the DM interaction on the domain wall motion

We get for the lattice point i and the neighbors $i \pm 1$:

$$\frac{\mathbf{D}_{\text{DM}}}{2} \cdot [(\mathbf{S}_{i-1} \times \mathbf{S}_i) + (\mathbf{S}_i \times \mathbf{S}_{i+1})]$$

$$\approx \frac{\mathbf{D}_{\text{DM}}}{2} \cdot \left[\left(\mathbf{S}(\mathbf{r}) - a\frac{\partial \mathbf{S}(\mathbf{r})}{\partial \mathbf{r}} \right) \times \mathbf{S}(\mathbf{r}) + \mathbf{S}(\mathbf{r}) \times \left(\mathbf{S}(\mathbf{r}) + a\frac{\partial \mathbf{S}(\mathbf{r})}{\partial \mathbf{r}} \right) \right]$$

$$= a\mathbf{D}_{\text{DM}} \cdot \left(\mathbf{S}(\mathbf{r}) \times \frac{\partial \mathbf{S}(\mathbf{r})}{\partial \mathbf{r}} \right)$$

Under the assumption $|\mathbf{S}(\mathbf{r})| = S$, which means that $\mathbf{S} = S\mathbf{e}_S$, we obtain for the DMI energy density:

$$\mathcal{E}_{\text{DM}} = aS^2 \, \mathbf{D}_{\text{DM}} \cdot \left(\mathbf{e}_S \times \frac{\partial \mathbf{e}_S}{\partial \mathbf{r}} \right) \tag{4.104}$$

With $\mathbf{D}_{\text{DM}} = D_{\text{DM}}\hat{\mathbf{z}}$, $\mathbf{e}_S = \begin{pmatrix} \sin\theta \cos\phi \\ \sin\theta \sin\phi \\ \cos\theta \end{pmatrix}$, as well as $\theta = \theta(\mathbf{r})$ and $\phi = \phi(\mathbf{r})$ we get:

$$\mathbf{D}_{\text{DM}} \cdot \left(\mathbf{e}_S \times \frac{\partial \mathbf{e}_S}{\partial \mathbf{r}} \right) = D_{\text{DM}} \left[(\sin\theta\cos\phi) \frac{\partial}{\partial \mathbf{r}} (\sin\theta\sin\phi) - (\sin\theta\sin\phi) \frac{\partial}{\partial \mathbf{r}} (\sin\theta\cos\phi) \right]$$

$$= D_{\text{DM}} \sin^2\theta \frac{\partial \phi}{\partial \mathbf{r}} . \tag{4.105}$$

Therefore, we can write the DMI energy in spherical coordinates as:

$$E = \int \text{d}r \, D_{\text{DM}} \sin^2\theta \, \phi' , \tag{4.106}$$

where the shortcut ϕ' means $\phi' = \partial\phi/\partial \mathbf{r}$. This energy is the micromagnetic counterpart of the discrete Hamiltonian Eq. (4.101).

Now, we are able to describe the domain wall dynamics influenced by the Dzyaloshinsky-Moriya interaction. We will start with the scenario where the Dzyaloshinsky-Moriya vector $\mathbf{D}_{\text{DM}}$ is parallel to the easy axis.

DM vector parallel to the easy axis: static domain wall solution

This scenario has been described by Tretiakov et al. [54]. However, the description will not follow the complex draft given by Tretiakov et al.. Here, we will discuss a less complex description using the q-ϕ model (see Sec. 4.2). This description is less confusing and gives a deeper understanding of the underlying physics. Furthermore, we will compare the analytical results with simulations to prove the accuracy of the analytical description. This has not been done by Tretiakov et al., which have only presented the analytical results.

The domain wall shall be located in a spin chain along the z-axis (see Fig. 4.17). The spins are oriented in $\pm z$-direction (easy axis in z direction). This means that the spins are oriented in the same direction as the spin chain (z-direction). Therefore, we are dealing with a transverse domain wall and not with a Bloch or Néel wall. Then, the energy with DMI is given by:

$$E = \int dz \left\{ A \left[\left(\frac{\partial \theta}{\partial z}\right)^2 + \sin^2\theta \left(\frac{\partial \phi}{\partial z}\right)^2 \right] - K\cos^2\theta \pm D_{\text{DM}} \sin^2\theta \frac{\partial \phi}{\partial z} \right\} . \tag{4.107}$$

Here, we assume that the DMI is parallel to the easy axis anisotropy: $\mathbf{D}_{\text{DM}} = D_{\text{DM}}\hat{\mathbf{z}}$. The other two terms in Eq. (4.107) are the exchange interaction A along the z-axis and an easy axis anisotropy in z-direction.

The calculation gets easier if we use dimensionless coordinates: $\mathcal{D} = D_{\text{DM}}/\sqrt{AK}$, $\tilde{E} = E/\sqrt{AK}$, $\tilde{z} = z\sqrt{K/A}$, ϕ, and θ are dimensionless. The energy E has the dimension $\sqrt{AK}$ and z the dimension $\sqrt{A/K}$.

Within the dimensionless coordinates the energy Eq. (4.107) becomes:

$$\tilde{E} = \int d\tilde{z} \left\{ \left[\left(\frac{\partial \theta}{\partial \tilde{z}}\right)^2 + \sin^2\theta \left(\frac{\partial \phi}{\partial \tilde{z}}\right)^2 \right] - \cos^2\theta \pm \mathcal{D}\sin^2\theta \frac{\partial \phi}{\partial \tilde{z}} \right\} . \tag{4.108}$$

For the Gilbert equation we then need the energy variations:

$$\frac{\delta \tilde{E}}{\delta \phi} = \frac{\partial \tilde{E}}{\partial \phi} - \frac{d}{d\tilde{z}}\frac{\partial \tilde{E}}{\partial \phi'} = 0 \tag{4.109}$$

$$\frac{\delta \tilde{E}}{\delta \theta} = \frac{\partial \tilde{E}}{\partial \theta} - \frac{d}{d\tilde{z}}\frac{\partial \tilde{E}}{\partial \theta'} = 0 , \tag{4.110}$$

with $\phi' = \partial \phi/\partial \tilde{z}$, and $\theta' = \partial \theta/\partial \tilde{z}$. For $\delta \tilde{E}/\delta \phi$ and $\delta \tilde{E}/\delta \phi$ we get:

$$\frac{\delta \tilde{E}}{\delta \phi} = -\frac{d}{d\tilde{z}}\left[\sin^2\theta \left(2\frac{\partial \phi}{\partial \tilde{z}} \pm \mathcal{D}\right)\right] = 0 \tag{4.111}$$

$$\frac{\delta \tilde{E}}{\delta \theta} = 2\sin\theta\cos\theta \left[\left(\frac{\partial \phi}{\partial \tilde{z}}\right)^2 + 1 \pm \mathcal{D}\frac{\partial \phi}{\partial \tilde{z}}\right] - 2\frac{\partial^2 \theta}{\partial \tilde{z}^2} = 0 . \tag{4.112}$$

The final goal is to solve these differential equations. From Eq. (4.111) we get immediately the non-trivial solution:

$$\frac{\partial \phi}{\partial \tilde{z}} = \mp \frac{\mathcal{D}}{2} . \tag{4.113}$$

4.6 Influence of the DM interaction on the domain wall motion

Inserting this in Eq. (4.112) leads to:

$$\frac{\partial^2 \theta}{\partial \tilde{z}^2} = \left[1 - \frac{\mathcal{D}^2}{4}\right] \sin\theta \cos\theta = K_{\text{eff}} \sin\theta \cos\theta . \tag{4.114}$$

This equation is similar to Eq. (4.6) for the transverse domain wall without Dzyaloshinsky-Moriya interaction:

$$A\frac{\partial^2 \theta}{\partial z^2} = K \sin\theta \cos\theta . \tag{4.115}$$

Following the standard procedure and multiplying both sides with $\partial\theta/\partial z$ and integrating over z leads to:

$$\left(\frac{\partial \theta}{\partial z}\right)^2 = K_{\text{eff}} \sin^2\theta + C . \tag{4.116}$$

C is the integration constant which is zero ($C = 0$) due to the boundary conditions at $z = \pm\infty$: $\theta = \text{const.} = 0$, resp. π and $\partial\theta/\partial z = 0$. The remaining differential equation (4.116) can be solved via the separation of variables. The result is:

$$\theta = 2\arctan\left(e^{\sqrt{K_{\text{eff}}}(\tilde{z}-\tilde{z}_0)}\right) . \tag{4.117}$$

Respectively using the original dimensions:

$$\theta = 2\arctan\left(e^{\frac{z-z_0}{\Delta}}\right) . \tag{4.118}$$

The domain wall width is given by:

$$\Delta = \sqrt{\frac{A}{K - \frac{D_{\text{DM}}^2}{4A}}} . \tag{4.119}$$

So far we have found the condition for $\delta\tilde{E}/\delta\theta$ by solving Eq. (4.112). To make the description unique we have to look for the condition for $\delta\tilde{E}/\delta\phi$. We can do this by solving equation (4.113). The result in dimensionless coordinates is given by:

$$\phi - \phi_0 = \mp\frac{\mathcal{D}}{2}(\tilde{z} - \tilde{z}_0) , \tag{4.120}$$

and therefore the corresponding solution with the correct dimensions becomes:

$$\phi - \phi_0 = \mp\Gamma(z - z_0) , \tag{4.121}$$

4 Field and current driven domain wall motion

with Γ given by:

$$\Gamma = \frac{D_{\text{DM}}}{2A} .\tag{4.122}$$

With this information, we are able to write down the domain wall profile as:

$$S_x = \sin\theta\cos\phi = \frac{\cos(\mp\Gamma(z-z_0)+\phi_0)}{\cosh\left(\frac{z-z_0}{\Delta}\right)} \tag{4.123}$$

$$S_y = \sin\theta\sin\phi = \frac{\sin(\mp\Gamma(z-z_0)+\phi_0)}{\cosh\left(\frac{z-z_0}{\Delta}\right)} \tag{4.124}$$

$$S_z = \cos\theta = \pm\tanh\left(\frac{z-z_0}{\Delta}\right) . \tag{4.125}$$

It can be seen that the domain wall is twisted due to the Dzyaloshinsky-Moriya

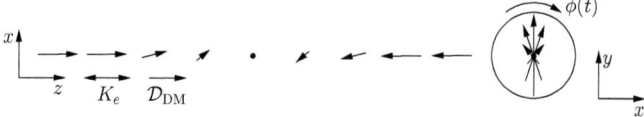

Figure 4.17: Sketch of the domain wall profile of a 1D transverse domain wall with Dzyaloshinsky-Moriya interaction: left hand side $x-z$ plane, right hand side (circle) $x-y$ plane. The DM vector $\mathcal{D}_{\text{DM}}$ is oriented parallel to the easy axis anisotropy K_e. Due to the Dzyaloshinsky-Moriya interaction and the absence of the hard axis anisotropy K_h the domain wall profile shows a twist. In this case we assume a precessional motion $d\phi/dt \neq 0$.

interaction (see Fig. 4.17). The domains themselves are not influenced, in the limit $z \to \pm\infty$ the normalized magnetization $\mathbf{S}(z)$ is given by:

$$S_x = S_y = 0 \tag{4.126}$$
$$S_z = \pm 1 . \tag{4.127}$$

DM vector parallel to the easy axis: domain wall dynamics

Until now, we have derived the static domain wall solution in a system with Dzyaloshinsky-Moriya interaction where the DM vector is pointing parallel to the

4.6 Influence of the DM interaction on the domain wall motion

easy axis. However, we are interested in the dynamics. The easiest way is to use the q-ϕ model introduced before.

We know that θ and ϕ are given by:

$$\theta(z,t) = 2\arctan\left(e^{-\frac{z-q(t)}{\Delta}}\right), \tag{4.128}$$

and

$$\phi(z,t) = \mp\Gamma\left(z - q(t)\right) + \phi_0. \tag{4.129}$$

Here, we have replaced $z_0(t)$ by $q(t)$.

The description which now follows is identical to the description given before in the case of the domain wall motion without Dzyaloshinsky-Moriya interaction. The starting point are the conditions:

$$\frac{\partial \theta}{\partial z} = -\frac{\sin \theta}{\Delta}, \tag{4.130}$$

$$\frac{\partial^2 \theta}{\partial z^2} = -\frac{\cos\theta \sin\theta}{\Delta^2}, \tag{4.131}$$

$$v = \dot{q} = \frac{\Delta}{\sin\theta}\dot{\theta}. \tag{4.132}$$

and

$$\frac{\partial \phi}{\partial z} = \frac{\partial}{\partial z}\left[\mp\Gamma\left(z-q(t)\right)+\phi_0\right] = \mp\Gamma, \tag{4.133}$$

with $q(t) = z_0$ and $\Gamma = \frac{D_{\text{DM}}}{A}$. It is quite easy to see that these conditions are fulfilled.

Then, the Gilbert equation in spherical coordinates ($S = \text{const.}$) is given by the following two differential equations for θ and ϕ:

$$\dot{\theta} = \frac{\gamma}{M_S \sin\theta}\frac{\delta\mathcal{E}}{\delta\phi} - \alpha\sin\theta\dot{\phi} - u_z\frac{\partial\theta}{\partial z} - \beta u_z \sin\theta\frac{\partial\phi}{\partial z} \tag{4.134}$$

$$\dot{\phi} = -\frac{\gamma}{M_S \sin\theta}\frac{\delta\mathcal{E}}{\delta\theta} + \frac{\alpha\dot{\theta}}{\sin\theta} - u_z\frac{\partial\phi}{\partial z} + \frac{\beta u_z}{\sin\theta}\frac{\partial\theta}{\partial z}. \tag{4.135}$$

The last two terms in both equations correspond to the spin torque terms and can be obtained from writing $d\mathbf{S}/dz$ and $\mathbf{S} \times d\mathbf{S}/dz$ in the Gilbert equation (3.109) in spherical coordinates ($S = \text{const.} = 1$):

$$\frac{d\mathbf{S}}{dz} = \frac{d\mathbf{e}_S}{dz} = \frac{\partial\theta}{\partial z}\mathbf{e}_\theta + \sin\theta\frac{\partial\phi}{\partial z}\mathbf{e}_\phi \tag{4.136}$$

4 Field and current driven domain wall motion

and

$$\mathbf{S} \times \frac{\mathrm{d}\mathbf{S}}{\mathrm{d}z} = (\mathbf{e}_S \times \mathbf{e}_\theta) \frac{\partial \theta}{\partial z} + \sin\theta \, (\mathbf{e}_S \times \mathbf{e}_\phi) \frac{\partial \phi}{\partial z} = \frac{\partial \theta}{\partial z} \mathbf{e}_\phi - \sin\theta \frac{\partial \phi}{\partial z} \mathbf{e}_\theta \,. \quad (4.137)$$

The energy $E = \int \mathrm{d}z \mathcal{E}$ is given by Eq. (4.107) plus the additional Zeeman term $\mathcal{E}_B = -M_S B \cos\theta$, which has not been taken into account in the static case. Then, the variations $\delta\mathcal{E}/\delta\phi$ and $\delta\mathcal{E}/\delta\theta$ are given (here in original dimensions):

$$\frac{\delta\mathcal{E}}{\delta\phi} = -2\sin\theta\cos\theta \left(\frac{\partial\theta}{\partial z}\right) \left(2A\frac{\partial\phi}{\partial z} \pm D_{\mathrm{DM}}\right) - \sin^2\theta \frac{\mathrm{d}}{\mathrm{d}z}\left(2\frac{\partial\phi}{\partial z} \pm D_{\mathrm{DM}}\right)$$
(4.138)

$$\frac{\delta\mathcal{E}}{\delta\theta} = 2\sin\theta\cos\theta \left[\left(A\frac{\partial\phi}{\partial z}\right)^2 + K \pm D_{\mathrm{DM}}\frac{\partial\phi}{\partial z}\right] - 2A\frac{\partial^2\theta}{\partial z^2} + M_S B \sin\theta \,.$$
(4.139)

The main assumption of the q-ϕ model is that we are in the center of the domain wall. Then, θ is given by $\theta = \pi/2$. Therefore, the variations become:

$$\frac{\delta\mathcal{E}}{\delta\phi} = \frac{\mathrm{d}}{\mathrm{d}z}\left(2\frac{\partial\phi}{\partial z} \pm D_{\mathrm{DM}}\right) = 0$$
(4.140)

$$\frac{\delta\mathcal{E}}{\delta\theta} = M_S B \,.$$
(4.141)

$\delta\mathcal{E}/\delta\phi$ becomes zero because of Eq. (4.133).

Then, we can write the Gilbert equations (4.134) and (4.135) using the conditions (4.130)-(4.133) and under the assumption of being in the center of the domain wall ($\theta = \pi/2$) as:

$$\dot{\theta} = -\alpha\dot{\phi} + \frac{u_z}{\Delta} \pm \beta u_z \Gamma$$
(4.142)

$$\dot{\phi} = \alpha\dot{\theta} - \gamma B \pm u_z \Gamma - \frac{\beta u_z}{\Delta} \,.$$
(4.143)

These differential equations can be solved exactly. The procedure is the same as before in section 4.2.

Inserting (4.143) in (4.142) and using the condition $v = \dot{\theta}\Delta$ we get:

$$v = \frac{\gamma B}{\alpha + \frac{1}{\alpha}}\Delta + \frac{1 + \alpha\beta}{1 + \alpha^2}u_z \mp \frac{\alpha - \beta}{1 + \alpha^2}\Gamma\Delta u_z \,.$$
(4.144)

This result is identical to the velocity of a field and current driven domain wall without Dzyaloshinsky-Moriya interaction Eq. (4.49), just with the additional

4.6 Influence of the DM interaction on the domain wall motion

term $\mp(\alpha - \beta)/(1 + \alpha^2)\Gamma \Delta u_z$ which describes the influence of the Dzyaloshinsky-Moriya interaction. This term increases / decreases the velocity depending on the sign of the DMI Eq. (4.101).

Inserting (4.142) in (4.143) leads to:

$$\dot{\phi} = \frac{\alpha - \beta}{1 + \alpha^2}\frac{u_z}{\Delta} - \frac{\gamma B}{1 + \alpha^2} \pm \frac{(1 + \alpha\beta)\Gamma}{1 + \alpha^2} u_z \, . \qquad (4.145)$$

Here, we can say the same as before in the case of the velocity: This is the result of a domain wall without Dzyaloshinsky-Moriya interaction Eq. (4.53) plus an additional term which modifies the result due to the DMI.

At this point it should be noticed that we have assumed to be in the center of the domain wall. This means we are in a moving frame. This is no problem for the domain wall velocity, because the velocity of the domain wall and the velocity of the moving frame are identical. However, due to the twist in the shape of the domain wall the moving frame also rotates. Therefore, if we want to investigate the oscillation of the domain wall we have to go into the stationary frame:

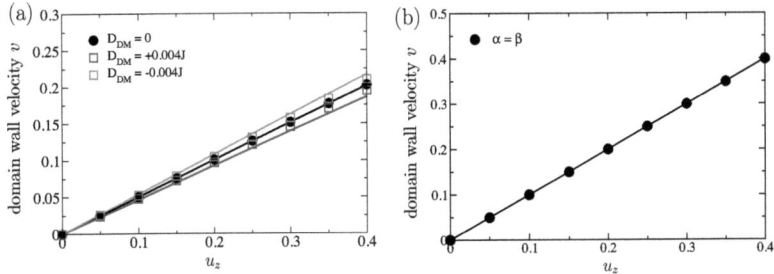

Figure 4.18: Domain wall velocity as function of u_z. (a) Depending on the sign and strength of the DMI we get an increase / decrease of the domain wall velocity. (b) Domain wall velocity in the case of $\alpha = \beta$. In this case the DMI has no effect.

We know that:

$$\frac{d\phi}{dz} = \mp\Gamma \Leftrightarrow d\phi = \mp\Gamma dz \, , \qquad (4.146)$$

4 Field and current driven domain wall motion

and therefore:

$$\dot{\phi} = \mp \Gamma \dot{z} \ . \tag{4.147}$$

This is the correction we have to take into account to change from the moving to the stationary frame. $v = \dot{z}$ is the velocity of the domain wall and at the same time that of the moving frame. Adding this term to Eq. (4.145) and taking into account Eq. (4.144) for the velocity, we get precession in the stationary frame:

$$\dot{\phi} = \frac{\alpha - \beta}{1 + \alpha^2} \left(\frac{1}{\Delta} \pm \Gamma^2 \Delta \right) u_z - \frac{\gamma B}{1 + \alpha^2} \left(1 \pm \alpha \Gamma \Delta \right) \ . \tag{4.148}$$

This means, we find for the precession velocity $v_{\text{prec.}}$:

$$v_{\text{prec.}} = \dot{\phi} \Delta = \frac{\alpha - \beta}{1 + \alpha^2} \left(1 \pm \Gamma^2 \Delta^2 \right) u_z - \frac{\gamma B \Delta}{1 + \alpha^2} \left(1 \pm \alpha \Gamma \Delta \right) \ . \tag{4.149}$$

In the limit $\mathbf{D}_{\text{DM}} = 0 \Leftrightarrow \Gamma = 0$ this equation is similar to Eq. (4.53).

Fig. 4.18 (a) shows the domain wall velocity as function of the current u_z ($B = 0$). The points correspond to the numerical and the lines to the analytical results. The change of the velocity due to the Dzyaloshinsky-Moriya interaction is clearly visible. Without DMI the velocity proposed by the analytical calculations and the velocity obtained by simulation show a perfect agreement. The velocity increases or decreases if the DMI has been taken into account. However, the analytical calculation overestimates the effect. The change of the velocity is smaller than expected from the analytical calculations. The reason for that can be searched in the assumptions made during the analytical calculations, especially for the current. These approximations represent an ideal situation and ignore that the current can change the shape of the domain wall which changes the dynamics. To get more realistic results we have to make more realistic assumptions, which however make analytical calculations impossible.

Fig. 4.18 (b) shows the situation of a current driven transverse domain wall when we assume $\alpha = \beta$. This is a situation which normally will not appear in reality. Nevertheless, in this case the Dzyaloshinsky-Moriya interaction has no effect. Here, we find again a perfect agreement between the analytical and the simulation results. The velocity in this scenario is as expected: $v = u_z$. Furthermore, the simulation shows that the transverse domain wall shows no precessional motion, which corresponds to $\dot{\phi} = 0$.

4.7 DM vector perpendicular to the easy axis anisotropy directions

The assumption of the previous section was that the DM vector is parallel to the easy axis anisotropy. In principle this has no necessary to be the case. Theoret-

4.7 DM vector perpendicular to the easy axis anisotropy directions

ically, the DM vector can show in any direction, depending on the symmetry of the system. To simplify the problem and due to the fact that we are interested in solvable analytical descriptions we restrict ourself in the following to the scenario that the DM vector is perpendicular to the easy axis anisotropy. Here, we investigate three scenarios: (1) the DM vector has an orientation perpendicular to the easy axis anisotropy K_e and parallel to the hard axis anisotropy K_h, (2) the DM vector is perpendicular to the easy as well hard axis anisotropy K_h, and (3) there is no additional hard axis anisotropy. In all these scenarios we assume that the spin chain has an alignment along the z axis, the easy axis is oriented in $\pm z$ direction (transverse domain wall) and the DM vector is oriented in y direction. To distinguish the three scenarios we have changed the strength (1),(2) $K_h \neq 0$, (3) $K_h = 0$ and orientation of the hard axis anisotropy K_h: (1) the hard axis anisotropy K_h is oriented in $\pm y$ direction, (2) in $\pm x$ direction.

The micromagnetic energy density $\mathcal{E}$ in the case of a DM vector in y direction parallel to the hard axis anisotropy K_h is given by:

$$\begin{aligned}\mathcal{E} &= A\left[\left(\frac{d\theta}{dz}\right)^2 + \sin^2\theta\left(\frac{d\phi}{dz}\right)^2\right] + \mathcal{D}_{\text{DM}}\cos\phi\frac{d\theta}{dz} \\ &- M_S B_z \cos\theta + K_h \sin^2\theta \sin^2\phi - K_e \cos^2\theta,\end{aligned}$$

For a hard axis anisotropy K_h in $\pm x$ direction we have to replace the term $K_h \sin^2\theta \sin^2\phi$ by $K_h \sin^2\theta \cos^2\phi$.

4.7.1 Direct Reversal: DM vector parallel to K_h

In this scenario the variation $\delta\mathcal{E}/\delta\theta$, with $B_z = 0$, leads to the following result:

$$\begin{aligned}\frac{\delta\mathcal{E}}{\delta\theta} &= (2A + \mathcal{D}_{\text{DM}}\cos\phi)\frac{d^2\theta}{dz^2} \\ &+ 2\sin\theta\cos\theta\left[K_e + K_h \sin^2\phi - A\left(\frac{d\phi}{dz}\right)^2\right] = 0\end{aligned} \quad (4.150)$$

With the assumption that ϕ is constant: $d\phi/dz = 0$ and $d\phi/dt = 0$ this differential equation can be easily solved and leads again to the well known domain wall profile (4.128) this time with the domain wall width:

$$\Delta = \sqrt{\frac{A + \frac{1}{2}\mathcal{D}_{\text{DM}}\cos\phi}{K_e + K_h \sin^2\phi}}. \quad (4.151)$$

The second variation $\delta\mathcal{E}/\delta\phi$, under the assumption to be in the center of the

4 Field and current driven domain wall motion

Figure 4.19: Sketch of the domain wall profile of a 1D transverse domain wall with Dzyaloshinsky-Moriya interaction: left hand side $x - z$ plane, right hand side (circle) $x - y$ plane. The DM vector $\mathcal{D}_{\text{DM}}$ is oriented perpendicular to the easy axis K_e and parallel to the hard axis anisotropy K_h. The Dzyaloshinsky-Moriya interaction only affects the domain wall width but not the profile itself. Due to the hard axis anisotropy K_h we can assume a direct reversal $\mathrm{d}\phi/\mathrm{d}t = 0$.

domain wall: $\theta = \pi/2$ and constant ϕ, leads to:

$$\frac{\delta \mathcal{E}}{\delta \phi} = \left(2K_h \cos\phi + \frac{\mathcal{D}_{\text{DM}}}{\Delta} \right) \sin\phi . \tag{4.152}$$

Therefore, we are able to write the Gilbert equations (4.134) and (4.135) under the assumption to be in the center of the domain wall ($\theta = \pi/2$) together with the conditions (4.130)-(4.133) and $\dot{\theta} = v/\Delta$ as:

$$\frac{v}{\Delta} = \frac{\gamma}{M_S} \left(2K_h \cos\phi + \frac{\mathcal{D}_{\text{DM}}}{\Delta} \right) \sin\phi + \frac{u_z}{\Delta} \tag{4.153}$$

$$0 = \frac{\alpha v}{\Delta} - \gamma B_z - \frac{\beta u_z}{\Delta} . \tag{4.154}$$

After eliminating v we get the stability condition:

$$\sin(\phi) = \frac{\gamma B_z + (\beta - \alpha)\frac{u_z}{\Delta}}{\frac{\alpha\gamma}{M_S}\left(2K_h \cos\phi + \frac{\mathcal{D}_{\text{DM}}}{\Delta} \right)} . \tag{4.155}$$

Inserting this result in (4.153) gives the formula for the velocity:

$$v = \frac{\gamma B_z}{\alpha}\Delta + \frac{\beta}{\alpha}u_z , \tag{4.156}$$

which is identical to the result of a transverse domain wall with direct reversal and no DMI [66, 52]. The only difference to the description without DMI is the stability criteria (4.155).

4.7 DM vector perpendicular to the easy axis anisotropy directions

Figure 4.20: Sketch of the domain wall profile of a 1D transverse domain wall with Dzyaloshinsky-Moriya interaction: left hand side $x - z$ plane, right hand side (circle) $x - y$ plane. The DM vector $\mathcal{D}_{\mathrm{DM}}$ is oriented perpendicular to the easy axis K_e and hard axis anisotropy K_h. The Dzyaloshinsky-Moriya interaction shows no influence on the profile in this case. Due to the hard axis anisotropy K_h we can assume a direct reversal $\mathrm{d}\phi/\mathrm{d}t = 0$.

4.7.2 Direct Reversal: DM vector perpendicular to K_h

This scenario is solvable if we assume that both anisotropies are dominating with respect to DMI: $K_h \gg \mathcal{D}_{\mathrm{DM}}$, $K_e \gg \mathcal{D}_{\mathrm{DM}}$. In this case the DMI can be neglected and we find a normal transverse domain wall with the profile (4.128) and the domain wall width:

$$\Delta = \sqrt{\frac{A}{K_e + K_h \sin^2 \phi}}. \qquad (4.157)$$

Then, the stability is characterized by:

$$\phi = \frac{1}{2} \arcsin \left(\frac{M_S B_z + (\beta - \alpha) \frac{u_x M_S}{\gamma \Delta}}{\alpha K_h} \right). \qquad (4.158)$$

and the velocity of the domain wall described by:

$$v = \frac{\gamma B_z}{\alpha} \Delta + \frac{\beta}{\alpha} u_z. \qquad (4.159)$$

4.7.3 Precessional motion

In this subsection we assume that there is no additional hard axis anisotropy K_h. This means that the system is invariant under rotation around the z axis if we ignore the DMI. The missing hard axis anisotropy also means that we can expect a rotation of the domain wall during the motion: $\mathrm{d}\phi/\mathrm{d}t \neq 0$. Furthermore, we assume for simplicity that $\mathrm{d}\phi/\mathrm{d}z = 0$. This is definitively the case when the transversal component of the domain wall and the DM vector $\vec{D}_{\mathrm{DM}}$ are perpendicular as in Fig. 4.21. In this case the DMI assists the magnetization reversal in the $x-z$ plane which is described by the angle θ. The assumption $\mathrm{d}\phi/\mathrm{d}z = 0$ is not necessarily the case when $\vec{D}_{\mathrm{DM}}$ is parallel to the transversal component of the domain wall. Under these assumptions we find:

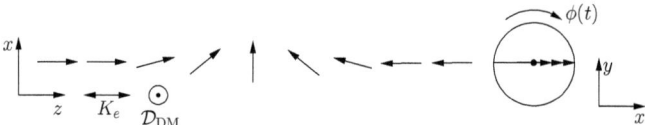

Figure 4.21: Sketch of the domain wall profile of a 1D transverse domain wall with Dzyaloshinsky-Moriya interaction: left hand side $x-z$ plane, right hand side (circle) $x-y$ plane. The DM vector $\mathcal{D}_{\mathrm{DM}}$ is oriented perpendicular to the easy axis anisotropy K_e. Due to the absence of the hard axis anisotropy K_h we can assume a precessional motion $\mathrm{d}\phi/\mathrm{d}t \neq 0$ which leads to a time dependent domain wall width $\Delta(t)$ and domain wall energy $E(t)$.

$$\frac{\delta \mathcal{E}}{\delta \theta} = (2A + \mathcal{D}_{\mathrm{DM}} \cos \phi) \frac{\mathrm{d}^2 \theta}{\mathrm{d}z^2} + 2K_e \sin\theta \cos\theta = 0 , \qquad (4.160)$$

the well know domain wall profile (4.128) together with the time dependent domain wall width:

$$\Delta(t) = \sqrt{\frac{A + \frac{1}{2}\mathcal{D}_{\mathrm{DM}} \cos \phi(t)}{K_e}} , \qquad (4.161)$$

and domain wall energy:

$$E(t) = 4\sqrt{\left(A + \frac{1}{2}\mathcal{D}_{\mathrm{DM}} \cos \phi(t)\right) K_e} . \qquad (4.162)$$

The second variation under the assumption to be in the center of the domain wall ($\theta = \pi/2$) leads to:

$$\frac{\delta \mathcal{E}}{\delta \phi} = \frac{\mathcal{D}_{\mathrm{DM}}}{\Delta} \sin \phi . \qquad (4.163)$$

4.7 DM vector perpendicular to the easy axis anisotropy directions

Then, it is easy to write the to write the Gilbert equations (4.134) and (4.135) with the assumption to be in the center of the domain wall ($\theta = \pi/2$) and with the conditions (4.130)-(4.133) as well as $\dot{\theta} = v/\Delta$ as:

$$\frac{v}{\Delta} = \alpha \frac{d\phi}{dt} - \frac{\gamma}{M_S} \frac{\mathcal{D}_{\text{DM}}}{\Delta} \sin\phi + \frac{u_z}{\Delta}, \quad (4.164)$$

and

$$\frac{d\phi}{dt} = \frac{\alpha v}{\Delta} - \gamma B_z - \frac{\beta u_z}{\Delta}. \quad (4.165)$$

Eliminating $d\phi/dt$ leads to the following velocity equation:

$$v(t) = \frac{\gamma B_z}{\alpha + \frac{1}{\alpha}} \Delta(t) + \frac{1+\alpha\beta}{1+\alpha^2} u_x - \frac{\gamma \mathcal{D}_{\text{DM}} \sin\phi(t)}{M_S \Delta(t)(1+\alpha^2)}. \quad (4.166)$$

In time average (mean value) the velocity of the domain wall is equal to:

$$\overline{v} = \frac{\gamma B_z}{\alpha + \frac{1}{\alpha}} \overline{\Delta} + \frac{1+\alpha\beta}{1+\alpha^2} u_x, \quad (4.167)$$

with the time averaged domain wall width:

$$\overline{\Delta} = \frac{1}{2} \left(\sqrt{\frac{A + \frac{1}{2}\mathcal{D}_{\text{DM}}}{K_e}} + \sqrt{\frac{A - \frac{1}{2}\mathcal{D}_{\text{DM}}}{K_e}} \right) \approx \sqrt{\frac{A}{K_e}}. \quad (4.168)$$

The approximation is correct if we assume a small DMI: $\mathcal{D}_{\text{DM}} \ll A$. In these cases the domain wall behaves in time average like a domain wall without DMI [66].

Eliminating v in (4.164) and (4.165) lead to the following differential equation for ϕ:

$$\frac{d\phi}{dt} = \frac{\alpha - \beta}{1+\alpha^2} \frac{u_x}{\Delta(t)} - \frac{\gamma B_z}{1+\alpha^2} - \frac{\alpha \gamma \mathcal{D}_{\text{DM}}}{M_S \Delta(t)(1+\alpha^2)} \sin\phi(t). \quad (4.169)$$

The solution of this differential equation is needed to calculate the velocity $v(t)$. However, this differential equation is quite complex and cannot be solved analytical in a complete way. However, to get an first impression we can make the assumption that the domain wall width is time-independent. In other words we make the following approximation: we replace $\Delta(t)$ by $\overline{\Delta}$. This approximation can be justified if $A \gg \mathcal{D}_{\text{DM}}$. In this case the differential equation becomes:

$$\frac{d\phi}{dt} = \phi_0 - \phi_1 \sin\phi(t), \quad (4.170)$$

4 Field and current driven domain wall motion

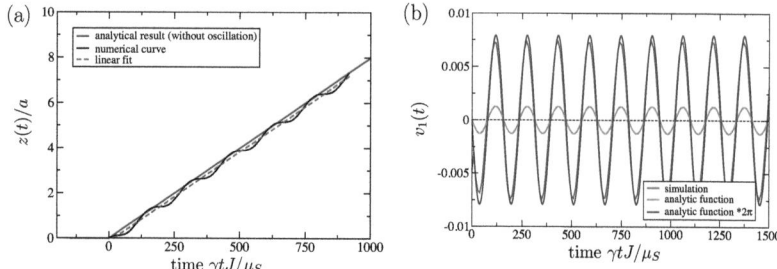

Figure 4.22: Domain wall motion of a transverse domain wall with Dzyaloshinsky-Moriya interaction: domain wall position as function of time corresponding to the scenario that the DM vector $\mathcal{D}_{\text{DM}}$ shows an orientation perpendicular to the easy axis anisotropy K_e and the assumption that there is no additional hard axis anisotropy $K_h = 0$. The last assumption leads to a precessional motion which leads to the oscillation in $z(t)$. The curves coming from computer simulations with: $\mu_S B_z/J = 0.2$, $D_{\text{DM}}/J = 0.004$, and $D_e/J = 0.005$.

with

$$\phi_0 = \frac{\alpha - \beta}{1+\alpha^2}\frac{u_x}{\overline{\Delta}} - \frac{\gamma B_z}{1+\alpha^2}, \qquad (4.171)$$

and

$$\phi_1 = \frac{\alpha\gamma\mathcal{D}_{\text{DM}}}{M_S\overline{\Delta}(1+\alpha^2)}, \qquad (4.172)$$

and can be solved easily by separation of variables. The result is:

$$\phi(t) = 2\arctan\left[\frac{\sqrt{\phi_0^2 - \phi_1^2}}{\phi_1}\tan\left(\frac{\sqrt{\phi_0^2 - \phi_1^2}\,t}{2}\right) - \frac{\phi_0}{\phi_1}\right]. \qquad (4.173)$$

An interesting situation appears if the domain wall is driven by an electrical current ($u_z \neq 0, B_z = 0$) and $\alpha = \beta$. It is known that for a tranverse domain wall without DMI ($\mathcal{D}_{\text{DM}}$) the domain wall shows no precession during the motion even if the domain wall normally shows a precessional motion for $\alpha \neq \beta$. If the domain wall configuration is like in Fig. 4.21: the DM vector is perpendicular to the domain wall plane the DMI will not force a precession during the motion [66]. In this case the solution of Eq. (4.170) is trivial: $d\phi/dt = 0 \Leftrightarrow \phi = \text{const.}$, and the velocity of the domain wall is given by:

$$v = u_x. \qquad (4.174)$$

4.7 DM vector perpendicular to the easy axis anisotropy directions

Fig. 4.22(a) shows the domain wall displacement as function of time for $\alpha < \beta$. The domain wall moves with the averaged velocity given by Eq. (4.167), however due to the oscillation coming from the DMI the domain wall arrives the end of the system with a delay or earlier depending on the sign of the DM vector $\mathcal{D}_{\text{DM}}$. Fig. 4.22 (b) shows the comparison between the oscillation which can be extracted form the simulations and the analytical result described by Eq. (4.166) and (4.170). $v_1(t)$ is given by the formula:

$$v_1(t) = \frac{\gamma \mathcal{D}_{\text{DM}} \sin \phi(t)}{M_S \Delta(t)(1+\alpha^2)}, \tag{4.175}$$

while for the analytical description $v_1(t)$ has to be assumed to be:

$$v_1^{\text{ana}}(t) = \frac{\gamma \mathcal{D}_{\text{DM}} \sin \phi(t)}{M_S \overline{\Delta}(1+\alpha^2)}. \tag{4.176}$$

And, $\phi(t)$ is described by Eq. (4.170).

It can be seen that the oscillation frequency is the same for the analytical as well as numerical result. However, the amplitude in the simulation is approximately six times larger than the analytical amplitude.

4 Field and current driven domain wall motion

5 Indirect manipulation of antiferromagnetic domain walls

The previous chapter has described the domain wall dynamics of ferromagnetic domain walls. Antiferromagnetic domain walls have not been discussed so far within this thesis. Classical antiferromagnets are characterized due to the existence of two sublattices with opposite spin direction. Therefore, it becomes clear that external fields cannot be used to drive an antiferromagnetic domain wall. If we want to control the antiferromagnetic domain wall motion we are restricted to the use of electric currents or stress effects. However, within this thesis these direct ways to manipulate an antiferromagnetic domain wall will not be discussed. This chapter describes two alternative ways which can be used to control the motion of an antiferromagnetic domain wall.

5.1 Antiferromagnetic domain wall motion in Exchange Bias systems

The motivation for considering antiferromagnetic domain wall motion in an Exchange Bias system is given by the Exchange Bias effect itself. Exchange Bias is a phenomenon in a FM/AFM layer system where the magnetization behavior of the antiferromagnet (AFM) causes a shift in the magnetization curve of the ferromagnet (FM). So far, the Exchange Bias effect is not well understood. There are different theories explaining the effect [56, 57], however none of them are generally accepted.

Experimental investigations of the domain structure of the ferromagnetic as well as the antiferromagnetic layer of an Exchange Bias system show identical domain structures of ferromagnet and antiferromagnet. It seems that not only the two layers are coupled but also the domain walls. Such a coupling has been shown experimentally by Metaxas et al. [59], however, in a ferromagnetic layer system.

The goal of the following study is to demonstrate the possibility to drive an antiferromagnetic domain wall in an Exchange Bias system via a ferromagnetic domain wall.

The starting conditions of this simulation are two relaxed domain walls, one in the antiferromagnetic layer at $x_i = 100a$, and one in the ferromagnetic layer at

5 Indirect manipulation of antiferromagnetic domain walls

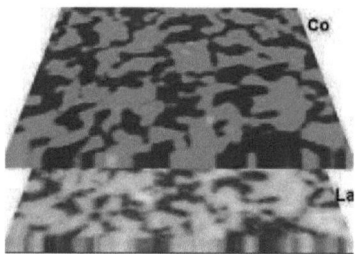

Figure 5.1: Exchange Bias system: thin layer of ferromagnetic cobalt grown on antiferromagnetic lanthanum iron oxide (LaFeO3). The ferromagnetic domain structure can be seen also in the antiferromagnet. To make the antiferromagnetic domains visible one has to rotate the spin orientation of one sublattice by 180° while the spins of the second sublattice are kept unchanged. The picture has been taken from the website of the Advanced Light Source Experimental System Group (ESG) at the Berkeley Lab [58]. The physics is described in [56, 57].

$x_n = 30a$ (see Fig. 5.2). The system is described by the following Hamiltonian:

$$\mathcal{H} = \mathcal{H}_{\text{AFM}} + \mathcal{H}_{\text{FM}} + \mathcal{H}_{\text{C}} \tag{5.1}$$

with

$$\mathcal{H}_{\text{AFM}} = J \sum_{\langle ij \rangle} \mathbf{S}_i \cdot \mathbf{S}_j - D_e \sum_i (S_i^x)^2 + D_h \sum_i (S_i^z)^2 \tag{5.2}$$

$$\mathcal{H}_{\text{FM}} = -J_{\text{FM}} \sum_{\langle nm \rangle} \mathbf{S}_n \cdot \mathbf{S}_m - d_e \sum_n (S_n^x)^2 + d_h \sum_n (S_n^z)^2 \tag{5.3}$$

$$\mathcal{H}_{\text{C}} = J_{\text{C}} \sum_{\langle in \rangle} \mathbf{S}_i \cdot \mathbf{S}_n \tag{5.4}$$

The Hamiltonian Eq. (5.1) contains three terms describing the antiferromagnet $\mathcal{H}_{\text{AFM}}$, the ferromagnet $\mathcal{H}_{\text{FM}}$, and the coupling between ferromagnetic and antiferromagnetic layer. The first term of $\mathcal{H}_{\text{AFM}}$ describes the nearest neighbor exchange interaction between the magnetic moments $S_i = \mu_i^{\text{AFM}}/\mu_s^{\text{AFM}}$ in the antiferromagnetic layer with $J = 1$. The second and third sum in $\mathcal{H}_{\text{AFM}}$ describe uniaxial anisotropies with an easy axis in x-direction ($D_e/J = 0.125$) and a hard axis in z-direction ($D_e/J = 0.05$).

5.1 Antiferromagnetic domain wall motion in Exchange Bias systems

The contributions of $\mathcal{H}_{\text{FM}}$ to the Hamiltonian $\mathcal{H}$ are similar to $\mathcal{H}_{\text{AFM}}$. The exchange constant here is $J_{\text{FM}} = J$ and the anisotropy constants for the easy axis and hard axis anisotropy are: $d_e/J = 0.075$ and $d_h/J = 4$.

The last contribution to $\mathcal{H}$ is $\mathcal{H}_{\text{C}}$. This term describes an antiferromagnetic exchange coupling between the magnetic moments of the antiferromagnet $\mathbf{S}_i$ and the magnetic moments of the ferromagnet $\mathbf{S}_n$. The coupling constant has been assumed to be $J_{\text{C}} = 0.5J$.

The fundamental equation of motion for the magnetic moments is the Landau-Lifshitz-Gilbert equation with additional spin torque terms:

$$\begin{aligned}
\frac{\mathrm{d}\mathbf{S}_i}{\mathrm{d}t} = & -\frac{\gamma}{\left(1+\alpha_i^2\right)\mu_S} \mathbf{S}_i \times \left[\mathbf{H}_i + \alpha_i \left(\mathbf{S}_i \times \mathbf{H}_i\right)\right] \\
& + -\frac{\alpha_i - \beta}{\left(1+\alpha_i^2\right)} u_x \mathbf{S}_i \times \frac{\mathrm{d}\mathbf{S}_n}{\mathrm{d}x} + \frac{1+\alpha_i\beta}{\left(1+\alpha_i^2\right)} u_x \mathbf{S}_i \times \left(\mathbf{S}_i \times \frac{\mathrm{d}\mathbf{S}_n}{\mathrm{d}x}\right)
\end{aligned} \quad (5.5)$$

Here, the index i corresponds to both subsystems: antiferromagnet as well as ferromagnet.

The spin torque terms describe the influence of the electric current, however, only in the ferromagnetic layer. Here, we assume an insulating antiferromagnetic layer and a conducting ferromagnet. Therefore, the electric current only flows through the ferromagnet and we have to assume $u_x = 0$ for the antiferromagnet and $u_x \neq 0$ and $\beta = 0.01$ in the case of the ferromagnet.

For the Gilbert damping the following assumptions have been made. The damping within the antiferromagnet is $\alpha_{\text{AFM}} = 0.025$ and within the ferromagnetic layer: $\alpha_{\text{FM}} = 0.02$.

Fig. 5.2b shows that both domain walls induce a magnetization in the y-direction in the opposite layer due to the interlayer coupling.

After switching on the current the ferromagnetic domain wall starts to move. A detailed description of the motion and corresponding velocities is given elsewhere [60]. When the ferromagnetic wall approaches the antiferromagnetic domain wall the ferromagnetic domain wall shifts the antiferromagnetic one, reducing the velocity of the ferromagnetic domain wall. The wall coupling and reduction of the domain wall velocity can be clearly seen in Fig. 5.3a. The velocity reduction can be explained by the fact that a domain wall can be seen as a quasi-particle of certain mass and the interaction process as an inelastic collision of two quasi-particles. In such a collision, if $D_z/J = 0$, the velocity of the ferromagnetic domain wall remains constant [see Fig. 5.3(b)]. In this case the velocity of the ferromagnetic domain wall is small (lower velocity limit, see [60]) and the antiferromagnetic domain wall has sufficient time to respond. However, regardless of the value of D_z/J, the speed of the antiferromagnetic and ferromagnetic domain wall is iden-

5 Indirect manipulation of antiferromagnetic domain walls

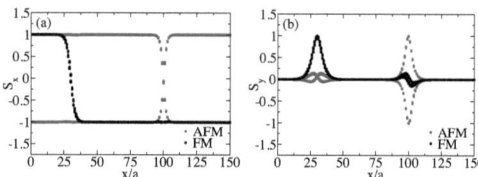

Figure 5.2: Domain wall profiles (a) easy axis direction and (b) transverse component of the ferromagnetic and antiferromagnetic domain wall before coupling.

tical and the distance between both domain walls stays the same, as is apparent in Fig. 5.3a-b.

Fig. 5.3c shows the y-component of the magnetization in both layers during the domain wall motion. It can be seen that the ferromagnetic domain wall is just behind the antiferromagnetic domain wall. It is remarkable that the center of the ferromagnetic domain wall is at the same place as the zero-crossing of the induced signal in the antiferromagnetic layer. The same is true for the signal of the antiferromagnetic domain wall in the ferromagnet. As can be seen in Fig. 5.3a after collision there is no change in the distance between the two domain walls, meaning the coupling is stable. However, if the ferromagnetic domain wall is faster than the highest possible velocity of the antiferromagnetic domain wall, the linking does not take place and the antiferromagnetic domain wall will not move.

Another interesting fact for this system is that an antiferromagnetic domain wall without an additional hard axis anisotropy ($D_z/J = 0$) follows a precessional motion only if the driving ferromagnetic domain wall also precesses during motion, i.e. the ferromagnet also has no hard axis anisotropy ($d_z/J = 0$). This behavior is explained by the coupling between the antiferromagnet and the ferromagnet. There are no external forces acting on the antiferromagnetic domain wall. Due to the interlayer exchange coupling the collinear orientation of the transverse components of the ferro- and antiferromagnetic domain walls is energetically favorable. Therefore, if the ferromagnetic domain wall precesses the antiferromagnetic domain wall must also follow the precessional motion. Anyhow, the antiferromagnetic domain wall precesses only if the ferromagnetic domain wall is precessing. This is independent of whether the antiferromagnetic domain wall would normally precess or not.

In a biaxial ferromagnet ($d_z/J \neq 0$) with currents higher than some critical current the domain wall starts to oscillate. This behavior is called Walker breakdown

5.1 Antiferromagnetic domain wall motion in Exchange Bias systems

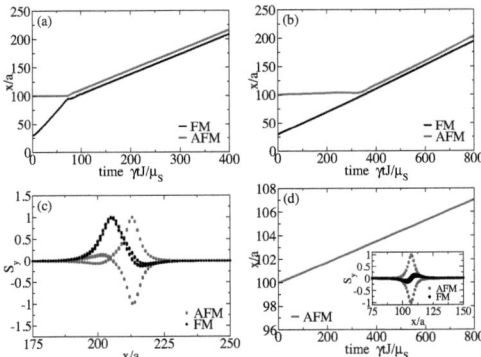

Figure 5.3: Domain wall displacement in a FM/AFM double layer: (a) with and (b) without hard axis anisotropy in the ferromagnetic layer. (c) transverse S_y components of the domain walls corresponding with Fig. 5.3(a) during coupled motion. (d) antiferromagnetic domain wall displacement without ferromagnetic domain wall.

[61]. In this regime the ferromagnetic domain wall starts to move periodically forward and backward. Due to the interlayer coupling the antiferromagnetic domain wall periodically moves forward and stops, depending on the distance between the two domain walls.

Thus far we have described the interaction between a ferromagnetic and an antiferromagnetic domain wall. However, in principle there is no need of a ferromagnetic domain wall to drive the antiferromagnetic domain wall. Fig. 5.3d shows the wall displacement of an antiferromagnetic domain wall driven by a current in the ferromagnetic layer. The domain wall is driven solely by the current acting on the magnetization component in the ferromagnet induced by the antiferromagnetic domain wall (see inset). The velocity in this case is very small but the motion is still present.

5.2 Antiferromagnetic domain walls manipulated with a SP-STM

Recent experiments [62, 63] have shown that it is possible to control the magnetization of a magnetic nanoisland by a spin polarized scanning tunneling microscope (SP-STM). The islands which have been investigated are iron on a tungsten (110) surface Fe/W(110). These nanoislands exhibit two stable magnetization directions due to a strong uniaxial anisotropy. The experiments have shown that it is possible to switch between these two states with current pulses coming from the tip of the spin-polarized scanning tunneling microscope. Furthermore, it could be demonstrated that the switching mechanism of the investigated nanoislands is given by the motion of one or more domain walls. Within the experiments it was not possible to answer the question about the number of domain walls and the detailed dynamics. However, corresponding simulations have demonstrated the correctness of the assumption that the reversal mechanism is described by the motion of domain walls. Fig. 5.4 presents a sequence of such a simulation

Figure 5.4: DW displacement of a ferromagnetic DW with a moving SP-STM tip. The tip is moving with a constant velocity marked by the dashed line. Depending on the orientation of the tip polarization **P** the DW will be shifted (pushed or pulled) by the tip or pushed away in the opposite direction.

which shows the creation and motion of a ring shaped domain wall created by the current coming from the tip of a spin-polarized scanning tunneling microscope. The simulations have shown that it must be possible to manipulate (push or pull)

5.2 Antiferromagnetic domain walls manipulated with a SP-STM

domain walls with a spin-polarized scanning tunneling microscope. Within this section we will show that the assumption that it must be possible to manipulate domain walls with an SP-STM is correct. The simulations describe transverse domain walls in a magnetic stripe manipulated by the current of an SP-STM. These simulations are similar to the simulation which have been performed to investigate the switching behavior of the Fe/W(110) nanoislands. Such a magnetic stripe can be seen as an analog of, e.g., a terrace on a stepped Fe/W(110) surface [64]. The only differences is that instead of a nanoisland a magnetic stripe has been investigated and the simulations have been performed with aid of a model Hamiltonian to reduce the numerical effort and to get more freedom in the description. Later, within this section I will show what happens if we use the correct material parameters for Fe/W(110) coming from *Ab-Initio* calculations.

The magnetic properties of the system are well described by the following Hamiltonian:

$$\mathcal{H} = -J\sum_{\langle ij \rangle} \mathbf{S}_i \cdot \mathbf{S}_j - K_x \sum_i (S_i^x)^2 + K_z \sum_i (S_i^z)^2 , \tag{5.6}$$

where $\mathbf{S}_i = \boldsymbol{\mu}_i/\mu_S$ is a three-dimensional magnetic moment of unit length. The first sum in Eq. (5.6) is the ferromagnetic exchange interaction between nearest neighbors. The last two terms represent uniaxial magnetocrystalline anisotropies, with the x-axis being the easy axis ($K_x > 0$) and the z-axis the hard axis ($K_z > 0$) of the system. We use two sets of constants which are on the order of a realistic material: set A: $J = 10$ meV, $K_x = 2.5$ meV and $K_z = 0.5$ meV, and set B: $J = 10$ meV, $K_x = 1.25$ meV and $K_z = 0.5$ meV. Set A gives a smaller DW which cannot be described by micromagnetism. Set B leads to a DW which can well be described by the analytical micromagnetism.

The underlying equation of motion for magnetic moments is the Landau-Lifshitz-Gilbert (LLG) equation with additional spin torque terms to describe the influence of the electric current:

$$\begin{aligned}\frac{\partial \mathbf{S}_i}{\partial t} =& -\frac{\gamma}{(1+\alpha^2)\mu_S} \mathbf{S}_i \times [\mathbf{H}_i + \alpha(\mathbf{S}_i \times \mathbf{H}_i)] \\ & - \mathcal{C}\mathbf{S}_i \times \mathbf{T}_i - \mathcal{D}\mathbf{S}_i \times (\mathbf{S}_i \times \mathbf{T}_i) ,\end{aligned} \tag{5.7}$$

with the gyromagnetic ratio γ, the Gilbert damping $\alpha = 0.025$ and internal field $\mathbf{H}_i = -\partial \mathcal{H}/\partial \mathbf{S}_i$. The last two terms are the two contributions of the spin torque [65, 66, 67, 68] with $\mathcal{C} = 0.05$ and $\mathcal{D} = 1$.

To describe the local strength and orientation of the current we follow the explanation of Tersoff and Hamann [69]:

$$\mathbf{T}_i = -I_0 e^{-2\kappa \sqrt{(x_i-x_{\text{tip}})^2 + (y_i-y_{\text{tip}})^2 + h^2}} \mathbf{P} . \tag{5.8}$$

5 Indirect manipulation of antiferromagnetic domain walls

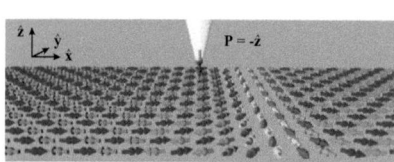

Figure 5.5: Manipulation of a ferromagnetic DW with a SP-STM tip.

$\kappa = 1.02 \cdot 10^{10}$ 1/m is the working function, $\mathbf{r}_{tip} = (x_{\text{tip}}, y_{\text{tip}}, h)$ is the time dependent tip position, $\mathbf{r}_i = (x_i, y_i, 1)$ the position of the spin $\mathbf{S}_i$, and $\mathbf{P}$ the tip polarization. $I_0 = 6 \cdot 10^{18}$ 1/s is the strength of the current. This value corresponds to a total current of approximately 0.6 μA, which is in the range of current strengths used in SP-STM experiments demonstrating current induced magnetization manipulation [63]. The tip itself moves with constant height of two lattice constants and a velocity of $v_{\text{tip}} = 1.2$ m/s along the long axis of the stripe (see Fig. 5.4). This velocity is much higher than the tip velocities in real experiments with max. $v_{\text{tip}} = 1000$ nm/s, but was necessary to achieve adequate computational time. However, our simulations show that it is possible to shift DW's also with such a tip velocity.

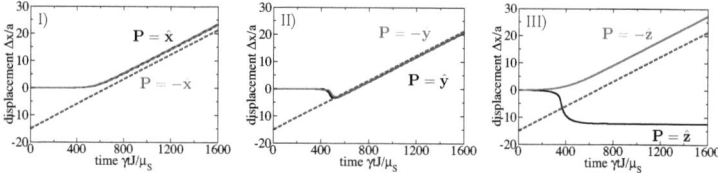

Figure 5.6: Domain wall displacement of an antiferromagnetic domain wall with a moving SP-STM tip. The dashed line describes the time evolution of the tip. The roman numbers correspond to different scenarios: I) tip polarization (anti)parallel to the domains, II) tip polarization (anti)parallel to the domain wall, and III) tip polarization perpendicular to domain and domain wall.

104

5.2 Antiferromagnetic domain walls manipulated with a SP-STM

The stripe contains a 180° transverse head-to-head DW separating two domains. Starting point of the tip in our simulation is fifteen lattice constants behind the center of the DW. Corresponding to different tip polarizations, different scenarios can be suggested: I) the tip polarization is parallel (D_p) or antiparallel (D_a) with respect to the magnetization of the domain underneath; II) parallel (W_p) or antiparallel W_a to the magnetization inside the DW or III) perpendicular (vertical) to both the magnetization inside the domains and DW. In this case the polarization can be the magnetization inside the DW rotated by ninety degree clockwise (V_{cw}) or counter-clockwise (V_{ccw}). In all scenarios we assume hundred percent tip polarization: $|\mathbf{P}| = 1$.

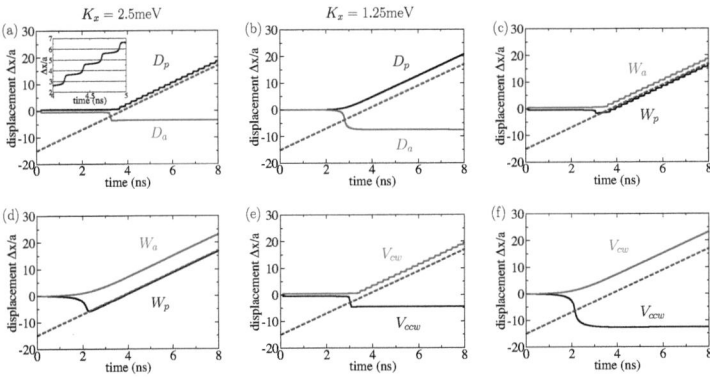

Figure 5.7: DW displacement of a ferromagnetic DW with a moving SP-STM tip. The tip is moving with a constant velocity marked by the dashed line. Depending on the orientation of the tip polarization **P** the DW will be shifted (pushed or pulled) by the tip or pushed away in the opposite direction.

The results of our simulation are shown in Fig. 5.6. It can be seen that it is possible to shift the domain wall. Depending on the tip polarization the domain wall will be pushed or pulled. In the case of a tip polarization which is parallel or antiparallel to the domains the domain wall will be pushed. This is independent if the tip polarization is in $\mathbf{P} = \hat{\mathbf{x}}$ or $\mathbf{P} = -\hat{\mathbf{x}}$. If the tip polarization is parallel or antiparallel to the domain wall the tip always pulls the domain wall independent

5 Indirect manipulation of antiferromagnetic domain walls

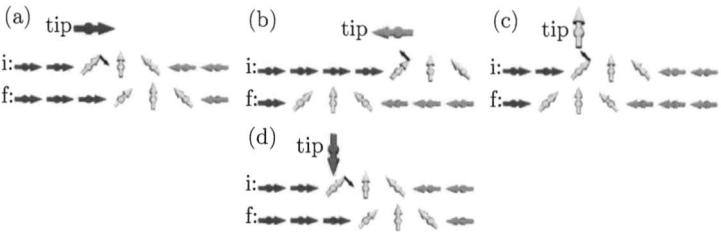

Figure 5.8: Explanation of the influence of a SP-STM tip on the DW corresponding to different tip polarizations (i: initial and f: final configuration).

of the two possible orientations of the tip: $\mathbf{P} = \pm\hat{\mathbf{x}}$. If the tip polarization is perpendicular to the domains and the domain wall, the wall will be pushed $\mathbf{P} = -\hat{\mathbf{z}}$ or pushed to the opposite direction backwards $\mathbf{P} = \hat{\mathbf{z}}$. In the latter case the tip looses the domain wall.

The explanation of the results is quite complicated, therefore we have also performed simulations with ferromagnetic domain walls. In this case the simulations become simpler and can be easily explained with some simple rules.

The results of these simulations are shown in Fig. 5.7. The different pictures show the DW displacement as well as the tip position (dashed lines) as a function of time. Fig. 5.7 (a),(c),(e) correspond to the parameter set A (narrow DW's), whereas Fig. 5.7 (b),(d),(f) correspond to the parameter set B (broad DW's). The response of both DW's with respect to the SP-STM tip is identical. It is clearly seen that the narrow DW shows a step-like motion caused by the high anisotropy. For the magnetization it is energetically more favorable to stay in the easy axis direction so that the magnetization does not follow the tip immediately but with a delay. The second difference is the averaged distance between tip and DW which correlates with the different DW widths.

Fig. 5.7(a) and (b) correspond to a magnetization parallel or antiparallel to the domain. D_p means that tip polarization and magnetization of the domain are parallel. In this situation the tip shifts the DW. D_a means that tip polarization and magnetization of the domain below are antiparallel. Now, the tip overtakes the DW and pushes it backwards. At the end the tip is above the opposite domain and magnetization of tip and domain are parallel.

Fig. 5.7(c) and (d) show the situations when the tip polarization is parallel or antiparallel with respect to the magnetization inside the domain wall. In the first case, W_p, we have a parallel alignment. The tip forces the DW to move towards

5.2 Antiferromagnetic domain walls manipulated with a SP-STM

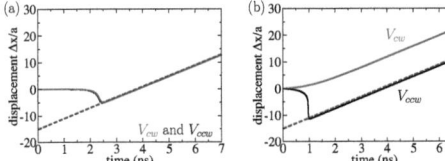

Figure 5.9: DW displacement of a ferromagnetic DW with a SP-STM tip. Here a modification of parameter set B and a tip polarization perpendicular to the magnetization of domains and DW has been used: (a) the assumed hard axis anisotropy has been set to zero, (b) the assumed current strength was set to $I_0 = 5 \cdot 10^{20} \frac{1}{s}$.

the tip position. Then the tip overtakes the DW and starts to pull the DW along the stripe. In the case of an antiparallel alignment W_a, the tip pushes the DW ahead along the stripe.

There are two simple rules. First rule: the system tends to stay in the energetically lowest configuration. The undistorted TDW is energetically more favorable than a bended DW. This is the main rule. The second rule is that the magnetization below the SP-STM tip tends to align parallel to the tip polarization if a spin current is flowing from the tip to the sample. This follows from the dominating last spin torque term of the LLG equation (5.7). Notice that the current is not strong enough to distort the magnetic moments inside the domains. A parallel alignment due to the current is possible only inside the DW where the spins have a reduced anisotropy energy and exchange coupling. In other words, the current does not influence the domains but the DW. An increased current would be necessary to manipulate also the magnetic moments inside the domains and to create a new domain.

With these rules we are able to understand the first two scenarios. Fig. 5.8 shows schematically what happens if the tip magnetization is parallel (a) or antiparallel (b) with respect to the magnetization inside the domains, or parallel (c) and antiparallel (d) to the magnetization inside the DW. The upper rows show the initial (i) and the lower rows the final (f) configurations, respectively. The small black arrow marks the acting spin torque $\mathbf{S}_i \times (\mathbf{S}_i \times \mathbf{T}_i)$. The tip is placed at the beginning of the DW. In the former three cases (a)-(c) the system follows directly the rules given before, especially, if the tip polarization is parallel to the magnetization inside the DW. In this case the DW will be caught by the tip first [see Fig. 5.8(c)] before both move together.

5 Indirect manipulation of antiferromagnetic domain walls

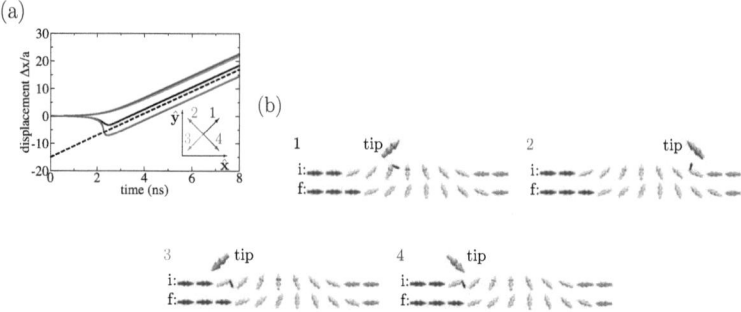

Figure 5.10: (a) DW displacement of a ferromagnetic DW with a moving SP-STM tip. The tip is moving with a constant velocity marked by the dashed line. The tip magnetization is rotated in-plane (xy-plane) by $(2n+1)\cdot 45°$, $n \in \{1,2,3,4\}$. (b) Explanation of the influence of a SP-STM tip on the DW corresponding to different tip polarizations corresponding with Fig. 5.10(a) (i: initial and f: final configuration).

In the case of an antiparallel alignment of tip and DW magnetization [see Fig. 5.8(d)] only the first rule will be fulfilled. The stable DW configuration will be reached with the lowest effort by pushing the DW. Now the tip magnetization and magnetic moment underneath have a perpendicular alignment. Furthermore, the magnetic moment below the tip is part of the domain now. As we have noticed, the current is not strong enough to distort the magnetization inside the domains. After turning the magnetic moment around 90°, it became part of the left domain. Therefore a further distortion is not possible. However, the more important first rule, the energy minimization, is fulfilled.

In 2D stripes the SP-STM tip is doing the same as described in these sketches here. It is amazing that the small spot size of the SP-STM tip is responsible for the displacement of the vast TDW. The spin current leads to a parallel alignment of the magnetic moments inside the spot below the tip. The relaxation to the energetically lower configuration leads to the displacement. The DW is thereby mostly undistorted and the transversal shape always perpendicular to the long stripe axis. This means we do not see any bending of the DW during the motion.

If the tip polarization is parallel to the direction of the hard axis anisotropy K_z, which means perpendicular to all magnetic moments in the system [scenario III, see Fig. 5.7(e) and (f)], the physics becomes quite complex. In this case the spin torque forces the magnetic moments in the direction of the hard axis anisotropy.

5.2 Antiferromagnetic domain walls manipulated with a SP-STM

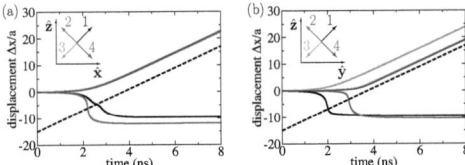

Figure 5.11: DW displacement of a ferromagnetic DW with a moving SP-STM tip. The tip is moving with a constant velocity marked by the dashed line. The tip magnetization is rotated out-of-plane (xz- resp. yz-plane) by $(2n+1) \cdot 45°$, $n \in \{1, 2, 3, 4\}$.

Figure 5.12: Reversal of the magnetization inside the domain wall using a SP-STM tip. The DW inside the DW appears as black spot.

This is energetically unfavorable and the magnetic moments evade this situation. The interplay between spin torque and the internal forces leads to a displacement of the DW forward V_{ccw} or backward V_{cw}. In the latter case, the tip will loose the DW, in the former case, the tip will shift the DW along the stripe. The direction of the displacement depends on the polarization of the tip, orientation of the magnetization inside the domain, the type of the transverse (head-to-head or tail-to-tail) DW, and the given sense of rotation of the vector product $\mathbf{S} \times \mathbf{H}_i$.

The situation becomes much easier if $K_z = 0$ or the current strength is strong enough to quasi neglect the energy barrier of the hard axis anisotropy. In this case the situation becomes identical with the situation described in Fig. 5.7(b): the tip pulls the DW and the magnetization below the tip is parallel to the tip polarization. Furthermore, in this case there is no difference between the two opposite polarizations V_{cw} and V_{ccw} (see Fig. 5.9).

5 Indirect manipulation of antiferromagnetic domain walls

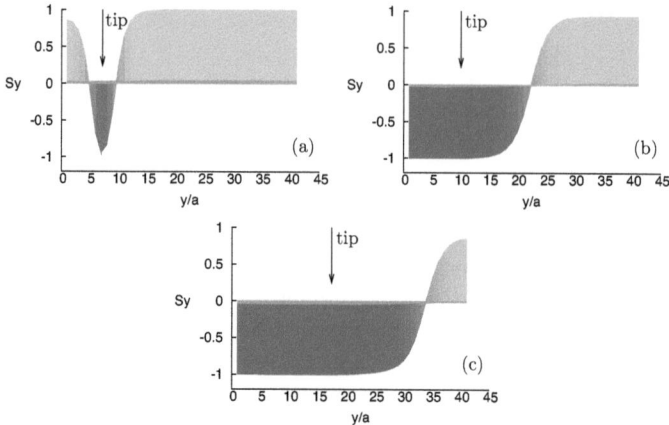

Figure 5.13: Reversal of the magnetization inside the domain wall (S_y) using a SP-STM tip (marked by the arrow). y gives the lateral position (width) within the wire in units of the lattice constant a. The figures (a), (b) and (c) show the situation at different times.

Up to now we have restricted our investigations to the three scenarios I-III corresponding to the three axes given by the cartesian coordinate system. In real experiments the orientation of the tip magnetization is more or less unknown. Furthermore, in almost all cases one can expect a tip magnetization not along one of these high symmetry axes [70]. In the following we investigate the situations of canted tip magnetizations. In these cases we can expect a superposition of the scenarios I, II, and III. Depending on the canting one scenario will dominate and the DW will show a similar response as in the case of the dominating scenario, however, modified by the other scenarios. To make the investigation more systematic we investigate the situations where the canting is 45° with respect to one of the symmetry axes.

Fig. 5.10(a) shows the DW displacement Δx of a TDW with paramter set B (broad DW's) and $I_0 = 6 \cdot 10^{18}$ 1/s manipulated by the current of a SP-STM tip with tip magnetization rotated by $\phi = (2n+1) \cdot 45°$, $n \in \mathbb{Z}$ in the film plane (xy-plane) with respect to the long stripe axis (x-axis). It can be seen that in all four cases the tip drives the DW. This is interesting because the tip coming from the left ($-x$) loses the DW if the tip magnetization is oriented in $-x$-direction. This situation is identical with $\mathbf{P} = \mathbf{D}_a$, the first scenario with antiparallel alignment

5.2 Antiferromagnetic domain walls manipulated with a SP-STM

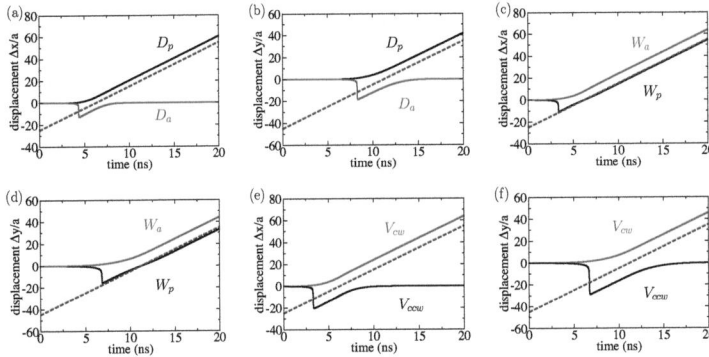

Figure 5.14: DW displacement in Fe/W(110) nanowires using a moving SP-STM tip: Fig. (a),(c) and (e) correspond to the left draft on top: DW oriented along [001] direction. Fig. (b),(d) and (f) correspond to the right draft: DW oriented along [1$\bar{1}$0] direction. The tip is moving with a constant velocity marked by the dashed line. Depending on the orientation of the tip polarization **P** the DW will be pushed or pulled by the tip.

of the magnetization of tip and domain underneath.

Fig. 5.10(b) gives the explanation for the behavior of the domain wall. To explain the response one can use the same rules as given before. The DW will stay in the energetically lowest configuration and the current will align the underlying magnetic moments. Therefore, it becomes clear that the situation 1 ($\phi = 45°$) is similar to the scenarios I or II with parallel orientation D_p, respectively W_p. Situation 2 ($\phi = 135°$) is similar to scenario II with parallel orientation W_p. However, the tip has to be in front of the domain wall. Situation 3 ($\phi = 225°$) is similar to the scenario II with antiparallel alignment W_a. Finally, situation 4 is similar to scenario I with parallel alignment D_p or scenerio II with antiparallel alignment W_a.

Fig. 5.11 shows the situation of a SP-STM tip canted out-of-plane (z-direction). Fig. 5.11(a) corresponds to a superposition of the scenarios I and III (rotation in the xz-plane), while Fig. 5.11(b) corresponds to a superposition of the scenarios II and III (rotation in the yz-plane). The canting angle is $\theta = (2n + 1) \cdot 45°$. Similar to scenario III the description is quite difficult. However, it can be seen that in both cases Fig. 5.11(a) and (b) the DW behaves like in the third scenario III: If the tip magnetization has a component in $+z$-direction corresponding to a counter-clockwise rotation of the tip magnetization with respect to the magnetization of the DW the tip will lose the DW. Otherwise, if the tip magnetization has a component in $-z$-direction corresponding to a clockwise rotation the tip pushes the DW. The canting of the tip magnetization only leads to a modification of scenario III. In other words, the out-of-plane component of the tip magnetization dominates and leads to a behavior similar to scenario III.

A similar behavior can be seen in Fig. 5.10. Here, the magnetization component of the SP-STM tip parallel, respectively antiparallel with respect to the magnetization of the DW dominates. At the end we can say: the vertical (out-of-plane) component of the tip-magnetization dominates all other components. The magnetization components parallel or antiparallel to the domains have the lowest influence.

By now we have reported the possibility to shift TDW's along the stripe. In principle it is also possible to switch the magnetization inside the DW using a SP-STM (see Fig. 5.12). Therefore, the tip has to be placed above the DW near one edge of the stripe. The tip polarization has to be opposite with respect to the magnetization inside the DW and one has to increase the current. As one can see, the tip creates a new domain [see Fig. 5.13(a)]. Then the tip has been moved along the transverse (y-) axis to the right. Thereby, the current shifts the DW inside the DW to the right stripe edge [see Fig. 5.12 and Fig. 5.13 (a)-(c)]. A small current unable to nucleate a new domain leads to the excitation of spin waves. These spin waves are captured inside the DW due to the reduced anisotropy energy [47]. A really strong current would lead to a magnetization reversal in combination with an excitation of the DW. This means the DW starts to move uncoordinated and the normal vector and shape of the DW are also changing accidentally.

So far we made the assumption of an ideal simple cubic system to explain the basics, but reality is often more complicated. Nevertheless, more complex simulations considering Fe/W(110) show similar results (see Fig. 5.14). The underlying material parameters have been taken from *Ab-Initio* calculations [71]: nearest neighbor exchange: $J_1 = 53.8$meV (FM), next nearest neighbor exchange: $J_2 = -13.5$meV (AFM), next next nearest neighbor exchange: $J_3 = 13.8$meV (FM), easy axis anisotropy $K_{[1\bar{1}0]} = -0.39$meV and hard axis anisotropy: $K_{[001]} = 2.42$meV.

The left column shows the simulation results corresponding to the three scenarios I-III and a stripe with long axis in $[1\bar{1}0]$ direction. The tip moves along the [1$\bar{1}$0] direction with a constant velocity and the DW is prolonged in [001] direction. The right column corresponds to a stripe along the [001] direction. The DW is oriented in [1$\bar{1}$0] direction and the tip moves along the [001] direction. In both cases the magnetization inside the domains is prolonged in [001], respectively [00$\bar{1}$], and inside the DW in [1$\bar{1}$0] direction (see drafts in Fig. 5.14). The change in the orientation of the DW leads to a change of the DW width [72]. We find for the DW prolonged in [001] direction (left column) a DW width of $\delta_{[001]} \approx 1.6$nm and for the DW prolonged in [1$\bar{1}$0] direction (right column) a width of $\delta_{[1\bar{1}0]} \approx 2.1$nm. Therefore, we have used different initial tip positions of 24, respectively 45 lattice constants behind the DW. The different domain wall widths can be seen also during the simulations as different distance between the tip and the DW during the motion (pushing).

The comparison between left and right column shows that even for the different DW orientations the results are identical. Furthermore, these results are similar to the results described before. However, there are some small differences. While in the previous system the tip loses the domain wall immediately in the cases D_a and V_{ccw}, the loss is delayed for Fe/W(110). The DW will be pulled for a while before it looses contact. This can be explained by the competing "long ranged" exchange interactions. The second difference is the interchange of V_{cw} and V_{ccw}. In the simple cubic system the tip pushes the DW if the tip polarization is V_{cw}, V_{ccw} leads to a loss. In Fig. 5.14 the DW will be pushed if V_{ccw}, and the tip looses contact if V_{cw}. This can be explained by the different rotation senses of the TDW. In the previous case the magnetization shows a left rotation while in the case of Fe/W(110) a right rotating DW was assumed. This shows again the complexity of scenario III.

The goal here was to investigate and describe the effects of spin torque coming from the current flow between magnetic STM tip and sample. To improve our model it is necessary to take into account Joule heating effects and the Oerstedt field in the simulation. Both are not included in our description. We think that the Joule heating helps to manipulate the DW's, while the Oerstedt field should not have a huge effect.

5 Indirect manipulation of antiferromagnetic domain walls

6 Summary

The goal of this thesis is to give an overview about the dynamics of transverse domain walls. Therefore, it is necessary to understand first the underlying Heisenberg model which has been introduced in the first chapter. Within this chapter we have seen that the Heisenberg model can be used to describe spin systems from a single atom up to macroscopic systems (micromagnetism). Here, we have to distinguish between the quantum mechanical and the classical Heisenberg model. The quantum mechanical Heisenberg model can be used to describe single atoms, small clusters, and molecules. The classical Heisenberg model can be used to describe systems on the atomic and macroscopic (micromagnetism) scale. Within this thesis the description has been restricted to the atomic scale and micromagnetic description and therefore, to the classical Heisenberg model. However, we have seen that the corresponding equation of motion, the Landau-Lifshitz-Gilbert equation can be derived from quantum mechanics.

Then, the Landau-Lifshitz-Gilbert equation has been used to describe the dynamics of domain walls. Here, we have restricted ourself to transverse domain walls. Most of the results we have got for the transverse domain walls can be used also to describe more complex domain walls like the vortex domain wall. Within this thesis the description has first been analytical using the θ-q-model. Later, the results have been proven with aid of numerical simulations to see the similarities and differences in both descriptions. The comparison shows that most of the analytical formulas give the correct results. However, if the situation becomes more complex the analytical description starts to become incorrect. Furthermore, the simulations show new effects like the appearance of spin waves during the domain wall motion. Such effects are mostly not included in the analytical descriptions. Spin waves behind the domain wall during the domain wall motion appear if a huge hard axis anisotropy prevents a Walker breakdown. In this case the domain wall cannot store all the energy gained during the domain wall motion due to the Zeeman term or electric current. The energy which cannot be stored in the domain wall anymore will be used to excite a spin wave wake behind the domain wall.

The last chapter has been used to discuss two possible ways to manipulate antiferromagnetic domain walls indirectly. The first way deals with an Exchange Bias system. Due to the coupling between ferromagnetic and antiferromagnetic

6 Summary

domain walls it is possible to push an antiferromagnetic domain wall with aid of a ferromagnetic domain wall. The ferromagnetic domain wall itself has been driven by a spin-polarized electric current which only flows through the ferromagnet. The second way is to use a spin-polarized scanning tunneling microscope to manipulate the domain wall. Here, we have seen that it is possible to shift domain walls, to excite spin waves, and to change the orientation of the magnetization inside the domain wall.

In summary this thesis provides an overview of the dynamics of domain walls.

A Appendix

A.1 Classical elliptical ferromagnetic spin waves (solving the Landau-Lifshitz equation)

Landau-Lifshitz equation (nearest neighbor interaction only):

$$\frac{d\boldsymbol{M}_n}{dt} = \frac{\gamma}{\mu_s}\left[J\boldsymbol{M}_n \times (\boldsymbol{M}_{n-1} + \boldsymbol{M}_{n+1}) + 2D_z\left(\boldsymbol{M}_n \times M_n^z \hat{\boldsymbol{z}}\right) - 2D_x\left(\boldsymbol{M}_n \times M_n^x \hat{\boldsymbol{x}}\right)\right] \quad (A.1)$$

Assumption: $M_z \approx +M$; $M_x, M_y \ll M \Rightarrow$

$$\frac{dM_n^x}{dt} \approx \frac{\gamma}{\mu_s}\left[-JM\left(M_{n-1}^y - 2M_n^y + M_{n+1}^y\right) + 2D_z M M_n^y\right]$$

$$\frac{dM_n^y}{dt} \approx \frac{\gamma}{\mu_s}\left[JM\left(M_{n-1}^x - 2M_n^x + M_{n+1}^x\right) - 2(D_z + D_x) M M_n^x\right] \quad (A.2)$$

$$\frac{dM_n^z}{dt} \approx 0$$

Solution ansatz:

$$M_n^x = u\cos(nka - \omega t)$$
$$M_n^y = v\sin(nka - \omega t)$$

$\Rightarrow$ with:

$$\cos((n\pm 1)ka - \omega t) = \cos(nka-\omega t)\cos(ka) \mp \sin(nka-\omega t)\sin(ka)$$
$$\sin((n\pm 1)ka - \omega t) = \sin(nka-\omega t)\cos(ka) \pm \cos(nka-\omega t)\sin(ka)$$

$$\omega u = \frac{2M\gamma}{\mu_s}\left[J(1-\cos(ka)) + D_z\right]v \quad (A.3)$$

$$\omega v = \frac{2M\gamma}{\mu_s}\left[J(1-\cos(ka)) + D_z + D_x\right]u \quad (A.4)$$

Solution if:

$$\begin{vmatrix} \omega & -\frac{2M\gamma}{\mu_s}\left[J(1-\cos(ka)) + D_z\right] \\ -\frac{2M\gamma}{\mu_s}\left[J(1-\cos(ka)) + D_z + D_x\right] & \omega \end{vmatrix} = 0$$

With $\gamma = \frac{g\mu_B}{\hbar}$ and $\mathcal{S} = \frac{g\mu_B}{\mu_s}M \Rightarrow$

$$\hbar\omega = 2\mathcal{S}\sqrt{\left[J(1-\cos(ka)) + D_z + D_x\right]\left[J(1-\cos(ka)) + D_z\right]} \quad . \quad (A.5)$$

A Appendix

A.2 Derivation of the Sine-Gordon equation

Energy with contributions coming from exchange (with exchange constant A), hard axis anisotropy (D, x-axis), easy axis (K, z-axis), and external field in $+z$-direction:

$$\mathcal{E} = A\left(\sin^2\theta\,(\partial_x\phi)^2 + (\partial_x\theta)^2\right) + D\sin^2\theta\cos^2\phi - K\cos^2\theta - SB\cos\theta \quad (A.6)$$

Coordinate system:

$$\mathbf{S} = S\begin{pmatrix} \sin\theta\cos\phi \\ \sin\theta\sin\phi \\ \cos\theta \end{pmatrix} \quad (A.7)$$

Derivation of the equation of motion (undamped Landau-Lifshitz equation):

$$\dot{\theta} = -\frac{\gamma}{S\sin\theta}\frac{\delta\mathcal{E}}{\delta\phi} \quad (A.8)$$

$$\dot{\phi} = \frac{\gamma}{S\sin\theta}\frac{\delta\mathcal{E}}{\delta\theta} \quad (A.9)$$

with $\dot{\phi} = \frac{\partial\phi}{\partial t}$ and $\dot{\theta} = \frac{\partial\theta}{\partial t}$. To get the equation of motion, we need the variation of $\mathcal{E}$ (Euler-Lagrange):

$$\frac{\delta\mathcal{E}}{\delta\phi} = \frac{\partial\mathcal{E}}{\partial\phi} - \frac{\partial}{\partial x}\left(\frac{\partial\mathcal{E}}{\partial\phi'}\right) \quad \text{with} \quad \phi' = \frac{\partial\phi}{\partial x} \quad (A.10)$$

$\frac{\partial\mathcal{E}}{\partial\theta}$ analog, replace ϕ by θ! The variations are given by:

$$\frac{\delta\mathcal{E}}{\delta\phi} = -2A\partial_x\left(\sin^2\theta\,\partial_x\phi\right) - 2D\sin^2\theta\cos\phi\sin\phi \quad (A.11)$$

$$\frac{\delta\mathcal{E}}{\delta\theta} = 2A\left[\sin\theta\cos\theta\,(\partial_x\phi)^2 - \partial_{xx}\theta\right] + 2D\sin\theta\cos\theta\cos^2\phi + 2K\cos\theta\sin\theta + SB\sin\theta. \quad (A.12)$$

With $\partial_x = \frac{\partial}{\partial x}$ and $\partial_{xx} = \frac{\partial^2}{\partial x^2}$, we get:

$$\dot{\theta} = \frac{2\gamma}{S\sin\theta}\left\{A\partial_x\left(\sin^2\theta\,\partial_x\phi\right) + D\sin^2\theta\cos\phi\sin\phi\right\} \quad (A.$$

$$\dot{\phi} = \frac{2\gamma}{S\sin\theta}\left\{A\left[\sin\theta\cos\theta\,(\partial_x\phi)^2 - \partial_{xx}\theta\right] + D\sin\theta\cos\theta\cos^2\phi + K\cos\theta\sin\theta + \frac{SB}{2}\sin\theta\right\} \quad (A.$$

A.2 Derivation of the Sine-Gordon equation

With additional Gilbert damping (Gilbert equation):

$$\dot{\theta} = \frac{2\gamma}{S\sin\theta}\left\{A\partial_x\left(\sin^2\theta\partial_x\phi\right) + D\sin^2\theta\cos\phi\sin\phi\right\} - \alpha\sin\theta\dot{\phi} \quad (A.15)$$

$$\dot{\phi} = \frac{2\gamma}{S\sin\theta}\left\{A\left[\sin\theta\cos\theta\left(\partial_x\phi\right)^2 - \partial_{xx}\theta\right] + D\sin\theta\cos\theta\cos^2\phi + K\cos\theta\sin\theta\right.$$

$$\left. + \frac{SB}{2}\sin\theta\right\} + \frac{\alpha\dot{\theta}}{\sin\theta} \quad (A.16)$$

Under the assumption $\phi = \frac{\pi}{2} + \epsilon$ with $\epsilon \ll 1$ we can write:

$$\sin\left(\frac{\pi}{2}+\epsilon\right) = \cos\epsilon \approx 1 \quad (A.17)$$

$$\cos\left(\frac{\pi}{2}+\epsilon\right) = -\sin\epsilon \approx -\epsilon \quad (A.18)$$

and

$$\dot{\theta} = \frac{2\gamma}{S\sin\theta}\left\{A\partial_x\left(\sin^2\theta\partial_x\epsilon\right) - D\sin^2\theta\epsilon\right\} - \alpha\sin\theta\dot{\epsilon} \quad (A.19)$$

$$\dot{\epsilon} = \frac{2\gamma}{S\sin\theta}\left\{A\left[\sin\theta\cos\theta\left(\partial_x\epsilon\right)^2 - \partial_{xx}\theta\right] + D\sin\theta\cos\theta\epsilon^2 + K\cos\theta\sin\theta\right.$$

$$\left. + \frac{SB}{2}\sin\theta\right\} + \frac{\alpha\dot{\theta}}{\sin\theta} \quad (A.20)$$

ϵ is a small number $\epsilon \ll 1$, therefore ϵ^2 is negligibly small. Therefore, the terms quadratic in ϵ can be skipped to first order. Furthermore, the approximation which leads to the double Sine-Gordon equation consists in discarding also the two terms $A\partial_x\left(\sin^2\theta\partial_x\epsilon\right)$ and $\alpha\sin\theta\dot{\epsilon}$ in the first equation because $D \gg 1$ is dominant with respect to A and α: $D \gg A$ as well as $D \gg \alpha$. With this approximation we get:

$$\dot{\theta} = -\frac{2D\gamma\sin\theta}{S}\epsilon \quad \Rightarrow \quad \epsilon\sin\theta = -\frac{S\dot{\theta}}{2D\gamma} \quad (A.21)$$

$$\dot{\epsilon}\sin\theta = \frac{\gamma}{S}\left\{-2A\partial_{xx}\theta + K\sin(2\theta) + SB\sin\theta\right\} + \alpha\dot{\theta}. \quad (A.22)$$

Here, we have used: $2\sin\theta\cos\theta = \sin(2\theta)$. Then, the time deriviative of $\sin\theta\epsilon$ is given by:

$$\frac{\partial}{\partial t}\left(\epsilon\sin\theta\right) = \epsilon\dot{\theta}\cos\theta + \dot{\epsilon}\sin\theta = -\frac{S\ddot{\theta}}{2D\gamma} \quad (A.23)$$

After Eq. (A.21) $\dot{\theta}$ is proportional to ϵ. Therefore, the first term of the middle part of Eq. (A.23) is quadratic in ϵ and can be skipped. The result is:

$$\dot{\epsilon}\sin\theta \approx -\frac{S\ddot{\theta}}{2D\gamma}. \quad (A.24)$$

A Appendix

Therefore, Eq. (A.24) becomes with Eq. (A.22):

$$-\frac{S\partial_{tt}\theta}{2D\gamma} \approx \frac{\gamma}{S}\{-2A\partial_{xx}\theta + K\sin(2\theta) + SB\sin\theta\} + \alpha\partial_t\theta . \tag{A.25}$$

Here, we have used the abbreviation $\partial_{tt}\theta = \frac{\partial^2\theta}{\partial t^2} = \ddot{\theta}$.

After reorganizing all terms we find the double Sine-Gordon equation:

$$\partial_{tt}\theta - \frac{4A\gamma^2 D}{S^2}\partial_{xx}\theta + \frac{2K\gamma^2 D}{S^2}\sin(2\theta) + \frac{2SB\gamma^2 D}{S^2}\sin\theta + \frac{2\alpha D\gamma}{S}\partial_t\theta = 0 . \tag{A.26}$$

With the abbreviations $\psi = 2\theta$, $c_0^2 = \frac{4A\gamma^2 D}{S^2}$, $\omega_0^2 = \frac{4K\gamma^2 D}{S^2}$, $f = \frac{4SB\gamma^2 D}{S^2}$, and $\Gamma = \frac{2\alpha\gamma D}{S}$ we get the final form of the double Sine-Gordon equation as given in section 4.3:

$$\partial_{tt}\psi - c_0^2\partial_{xx}\psi + \omega_0^2\sin\psi + f\sin\left(\frac{\psi}{2}\right) + \Gamma\partial_t\psi = 0 . \tag{A.27}$$

Without damping $\Gamma \propto \alpha$ and external field $f \propto B$, this equation reduces to the normal undamped Sine-Gordon equation:

$$\partial_{tt}\psi - c_0^2\partial_{xx}\psi + \omega_0^2\sin\psi = 0 . \tag{A.28}$$

Please notice:

$$\frac{c_0}{\omega_0} = \sqrt{\frac{A}{K}} \tag{A.29}$$

is the "domain wall width".

In section 3.3 we have introduced additional terms describing the influence of a spin-polarized electric current. In this case we have to add $-u_x\frac{\partial\theta}{\partial x}$ to Eq. (A.15) and $\frac{\beta u_x}{\sin\theta}\frac{\partial\theta}{\partial x}$ to Eq. (A.16). Then, the calculation is the same as described before. The new term $-u_x\frac{\partial\theta}{\partial x}$ in Eq. (A.15) will be skipped during the approximation because $D \gg u_x$. Finally, we get instead of Eq. (A.26):

$$\partial_{tt}\theta - \frac{4A\gamma^2 D}{S^2}\partial_{xx}\theta + \frac{2D\gamma\beta u_x}{S}\partial_x\theta + \frac{2K\gamma^2 D}{S^2}\sin(2\theta) + \frac{2SB\gamma^2 D}{S^2}\sin\theta + \frac{2\alpha D\gamma}{S}\partial_t\theta = 0 . \tag{A.30}$$

With the abbreviations: $\psi = 2\theta$, $c_0^2 = \frac{4A\gamma^2 D}{S^2}$, $\omega_0^2 = \frac{4K\gamma^2 D}{S^2}$, $f = \frac{4SB\gamma^2 D}{S^2}$, $\Gamma = \frac{2\alpha\gamma D}{S}$, and $u_0 = \frac{2D\gamma\beta u_x}{S}$ we finally get:

$$\partial_{tt}\psi - c_0^2\partial_{xx}\psi + u_0\partial_x\psi + \omega_0^2\sin\psi + f\sin\left(\frac{\psi}{2}\right) + \Gamma\partial_t\psi = 0 . \tag{A.31}$$

Bibliography

[1] J. M. D. Coey, *Magnetism and Magnetic Materials* (Cambridge University Press, Cambridge, 2009).

[2] S. J. Blundell, *Magnetism: A Very Short Introduction* (Oxford University Press, Oxford, 2012).

[3] M. Getzlaff, *Fundamentals in Magnetism* (Springer Verlag, Berlin, 2010).

[4] C. A. Ulrich, *Time-Dependent Density-Functional Theory* (Oxford University Press, Oxford, 2012).

[5] L. Piela, *Ideas of Quantum Chemistry* (Elsevier, Amsterdam, 2007).

[6] D. Hobbs und J. Hafner, *Fully unconstrained non-collinear magnetism in triangular Cr and Mn monolayers and overlayers on Cu(111) substrates*, J. Phys.: Cond. Matter **12**, 7025 (2000).

[7] A. J. Freeman und K. Nakamura, *Modern computational magnetism: role of noncollinear magnetism in complex magnetic phenomena*, phys. stat. sol. (b) **241**, 1399 (2004).

[8] J. D. Jackson, *Classical Electrodynamics* (John Wiley & Son, New York, 1999).

[9] K. Mølmer et al., *Monte Carlo wave-function method in quantum optics*, J. Opt. Soc. Am. B **10**, 524 (1993).

[10] N. Gisin, *Spin relaxation and dissipative Schrödinger like evolution equations*, Helv. Phys. Acta **54**, 457 (1981).

Bibliography

[11] M. Lakshmanan, *The fascinating world of the Landau-Lifshitz-Gilbert equation: an overview*, Phil. Trans. R. Soc. A **369**, 1280 (2011).

[12] C.-S. Liu, K.-C. Chen und C.-S. Yeh, *A mathematical revision of the Landau-Lifshitz equation*, J. of Marine Sci. and Tech. **17**, 228 (2009).

[13] E.-M. Graefe, M. Höning und H. J. Korsch, *Classical limit of non-Hermitian quantum dynamics - a generalised canonical structure*, J. Phys. A **43**, 075306 (2010).

[14] E. Schrödinger, *Die gegenwärtige Situation der Quantenmechanik*, Naturwissenschaften **23**, 807 (1935).

[15] R. Wieser, *Quantum Spin Dynamics*, arXiv, 1410.6383 (2014).

[16] R. Balakrishnan und A. R. Bishop, *Nonlinear dynamics of a quantum ferromanetic chain: Spin-coherent-state approach*, Phys. Rev. B **40**, 9194 (1989).

[17] R. Balakrishnan, J. A. Holyst und A. R. Bishops, *Soliton dynamics in the uniaxially anisotropic quantum ferromagnetic chain*, Journal of Physics: Condensed Matter **2**, 1869 (1990).

[18] J. M. Radcliffe, *Some properties of coherent spin states*, J. Phys. A: Gen. Phys. **4**, 313 (1971).

[19] R. Kikuchi, *On the Minimum of the Magnetization Reversal Time*, J. Appl. Phys. **27**, 1352 (1956).

[20] T. L. Gilbert und J. M. Kelly, *Proceedings of the Pittsburgh Conference on Magnetism and Magnetic Materials* (Am. Inst. Electr. Engrs., October, 1955), S. 253.

[21] T. L. Gilbert, *A phenomenological theory of damping in ferromagnetic materials*, IEEE Trans. Mag. **40**, 3443 (2004).

[22] D. Altwein, private communication.

[23] V. Weisskopf und E. Wigner, *Über die natürliche Linienbreite in der Strahlung des harmonischen Oszillators*, Z. Physik **65**, 18 (1930).

[24] R. Wiesendanger, *Scanning Probe Microscopy and Spectroscopy* (Cambridge University Press, Cambridge, 1994).

[25] C. Schieback, M. Kläui, U. Nowak, U. Rüdiger und P. Nielaba, *Numerical investigation of spin-torque using the Heisenberg model*, Eur. Phys. J. B **59**, 429 (2007).

[26] B. Krüger, Ph.D. thesis, Universität Hamburg, 2011.

[27] R. Wieser, U. Nowak und K. D. Usadel, *Domain wall mobility in nanowires: transverse vs. vortex walls*, Phys. Rev. B **69**, 064401 (2004).

[28] R. Wieser, U. Nowak und K. D. Usadel, *Domain wall motion in nanowires*, Phase Transit. **78**, 115 (2005).

[29] A. P. Malozemoff und J. C. Slonczewski, *Magnetic Domain Walls in Bubble Materials* (Academic Press, New York, 1979).

[30] W. Döring, *Über die Trägheit der Wände zwischen Weißschen Bezirken*, Z. Naturforschung **3a**, 373 (1948).

[31] M. Yan, A. Kákay, S. Gliga und R. Hertel, *Beating the Walker Limit with Massless Domain Walls in Cylindrical Nanowires*, Phys. Rev. Lett. **104**, 057201 (2010).

[32] H. J. Mikeska, *Solitons in one-dimensional magnets*, J. Appl. Phys. **52**, 1950 (1981).

[33] J. K. Kjems und M. Steiner, *Evidence for Soliton Modes in the One-Dimensional Ferromagnet $CsNiF_3$*, Phys. Rev. Lett. **41**, 1137 (1978).

Bibliography

[34] J. P. Boucher, L. P. Regnault, J. Rossat-Mignod, J. P. Renard, J. Bouillot und W. G. Stirling, *Solitons in the one-dimensional antiferromagnet TMMC*, Sol. State Comm. **33**, 171 (1980).

[35] H. Forster, T. Schrefl, D. Suess, W. Scholz, V. Tsiantos, R. Dittrich und J. Fidler, *Domain wall motion in nano-wires using moving grids*, J. Appl. Phys. **91**, 6914 (2002).

[36] R. Hertel und J. Kirschner, *Magnetization reversal dynamics in nickel nanowires*, Physica B **343**, 206 (2004).

[37] N. L. Schryer und L. R. Walker, *The motion of 180° domain walls in uniform dc magnetic fields*, J. Appl. Phys. **45**, 5406 (1974).

[38] J. F. Dillon, in *Magnetism*, hrsg. von G. T. Rado und H. Suhl (Academic Press, New York, 1963), Vol. 1, S. 149.

[39] D. L. Landau und E. M. Lifshitz, *On the theory of the dispersion of magnetic permeability in ferromagnetic bodies*, Phys. Z. Sowjetunion **8**, 153 (1935).

[40] D. G. Porter und M. J. Donahue, *Velocity of transverse domain wall motion along thin, narrow stripes*, J. Appl. Phys. **95**, 6729 (2004).

[41] H. Suhl, *Theory of the magnetic damping constant*, IEEE Trans. Mag. **34**, 1834 (1998).

[42] V. L. Safonov und H. N. Bertram, *Spin-wave dynamic magnetization reversal in a quasi-single-domain magnetic grain*, Phys. Rev. B **63**, 094419 (2001).

[43] R. W. Chantrell, J. D. Hannay, M. Wongsam, T. Schrefl und H.-J. Richter, *Computational approaches to thermally activated fast relaxation*, IEEE Trans. Mag. **34**, 1839 (1998).

[44] O. Chubykalo, J. D. Hannay, M. Wongsam, R. W. Chantrell und J. M. Gonzalez, *Langevin dynamic simulation of spin waves in a micromagnetic model*, Phys. Rev. B **65**, 184428 (2002).

[45] H. J. Mikeska, *Solitons in a one-dimensional magnet with an easy plane*, J. Phys. C **11**, L29 (1978).

[46] H. How, R. C. O'Handley und F. R. Morgenthaler, *Soliton theory for realistic magnetic domain-wall dynamics*, Phys. Rev. B **40**, 4808 (1989).

[47] R. Wieser, E. Y. Vedmedenko und R. Wiesendanger, *Quantized spin waves in ferromagnetic and antiferromagnetic structures with domain walls*, Phys. Rev. B **79**, 144412 (2009).

[48] S. S. P. Parkin, M. Hayashi und L. Thomas, *Magnetic Domain-Wall Racetrack Memory*, Science **320**, 190 (2008).

[49] S. E. Barnes und S. Maekawa, *Current-Spin Coupling for Ferromagnetic Domain Walls in Fine Wires*, Phys. Rev. Lett. **95**, 107204 (2005).

[50] S. Zhang und Z. Li, *Roles of Nonequilibrium Conduction Electrons on the Magnetization Dynamics of Ferromagnets*, Phys. Rev. Lett. **93**, 127204 (2004).

[51] M. E. Lucassen, H. J. van Driel, C. M. Smith und R. A. Duine, *Current-driven and field-driven domain walls at nonzero temperature*, Phys. Rev. B **79**, 224411 (2009).

[52] A. Mougin, M. Cormier, J. P. Adam, P. J. Metaxas und J. Ferrè, *Domain wall mobility, stability and Walker breakdown in magnetic nanowires*, Europhys. Lett. **78**, 57007 (2007).

Bibliography

[53] Z. Li und S. Zhang, *Domain-Wall Dynamics and Spin-Wave Excitations with Spin-Transfer Torques*, Phys. Rev. Lett. **92**, 207203 (2004).

[54] O. A. Tretiakov und A. Abanov, *Current Driven Magnetization Dynamics in Ferromagnetic Nanowires with a Dzyaloshinsky-Moriya Interaction*, Phys. Rev. Lett. **105**, 157201 (2010).

[55] A. Thiaville, S. Rohart, E. Jue, V. Cros und A. Fert, *Dynamics of Dzyaloshinskii domain walls in ultrathin magnetic films*, Europhys. Lett. **100**, 57002 (2012).

[56] F. Nolting et al., *Direct observation of the alignment of ferromagnetic spins by antiferromagnetic spins*, Nature **405**, 767 (2000).

[57] A. Scholl et al., *Observation of antiferromagnetic domains in epitaxial thin films*, Science **287**, 1014 (2000).

[58] http://xraysweb.lbl.gov/peem2/webpage/Research.shtml.

[59] P. J. Metaxas et al., *Dynamic Binding of Driven Interfaces in Coupled Ultrathin Ferromagnetic Layers*, Phys. Rev. Lett. **104**, 237206 (2010).

[60] R. Wieser et al., *Current-driven domain wall motion in cylindrical nanowires*, Phys. Rev. B **82**, 144430 (2010).

[61] A. Thiaville und Y. Nakatani, in *Spin dynamics in confined magnetic structures III*, hrsg. von B. Hillebrands und A. Thiaville (Springer, Berlin / Heidelberg, 2006).

[62] S. Krause, L. Berbil-Bautista, G. Herzog, M. Bode und R. Wiesendanger, *Current-Induced Magnetization Switching with a Spin-Polarized Scanning Tunneling Microscope*, Science **317**, 1537 (2007).

[63] G. Herzog, S. Krause und R. Wiesendanger, *Heat assisted spin torque switching of quasistable nanomagnets across a vacuum gap*, Appl. Phys. Lett. **96**, 102505 (2010).

[64] M. Pratzer und H. J. Elmers, *Domain wall energy in quasi-one-dimensional Fe/W(110) nanostripes*, Phys. Rev. B **67**, 094416 (2003).

[65] J. C. Slonczewski, *Current-driven excitation of magnetic multilayers*, J. Magn. Magn. Mat. **159**, L1 (1996).

[66] R. Wieser, E. Y. Vedmedenko, P. Weinberger und R. Wiesendanger, *Current-driven domain wall motion in cylindrical nanowires*, Phys. Rev. B **82**, 144430 (2010).

[67] Z. Li und S. Zhang, *Magnetization dynamics with a spin-transfer torque*, Phys. Rev. B **68**, 024404 (2003).

[68] D. V. Berkov und N. L. Gorn, *Spin-torque driven magnetization dynamics in a nanocontact setup for low external fields: Numerical simulation study*, Phys. Rev. B **80**, 064409 (2009).

[69] J. Tersoff und D. R. Hamann, *Theory and Application for the Scanning Tunneling Microscope*, Phys. Rev. Lett. **50**, 1998 (1983).

[70] A. Schlenhoff, S. Krause, G. Herzog und R. Wiesendanger, *Bulk Cr tips with full spatial magnetic sensitivity for spin-polarized scanning tunneling microscopy*, Appl. Phys. Lett. **97**, 083104 (2010).

[71] M. Heide, Ph.D. thesis, RWTH Aachen, 2006.

[72] E. Y. Vedmedenko, A. Kubetzka, K. von Bergmann, O. Pietzsch, M. Bode, J. Kirschner, H. P. Oepen und R. Wiesendanger, *Domain Wall Orientation in Magnetic Nanowires*, Phys. Rev. Lett. **92**, 077207 (2004).

i want morebooks!

Buy your books fast and straightforward online - at one of the world's fastest growing online book stores! Environmentally sound due to Print-on-Demand technologies.

Buy your books online at
www.get-morebooks.com

Kaufen Sie Ihre Bücher schnell und unkompliziert online – auf einer der am schnellsten wachsenden Buchhandelsplattformen weltweit!
Dank Print-On-Demand umwelt- und ressourcenschonend produziert.

Bücher schneller online kaufen
www.morebooks.de

OmniScriptum Marketing DEU GmbH
Heinrich-Böcking-Str. 6-8
D - 66121 Saarbrücken
Telefax: +49 681 93 81 567-9

info@omniscriptum.de
www.omniscriptum.de

Printed by Books on Demand GmbH, Norderstedt / Germany